The Arctic and Environmental Change

The Arctic and Environmental Change

EDITED BY

P. WADHAMS

Scott Polar Research Institute, University of Cambridge, UK

J. A. DOWDESWELL

Centre for Glaciology, Institute of Earth Studies, University of Wales, UK

A. N. SCHOFIELD

Department of Engineering, University of Cambridge, UK

CRC Press
Taylor & Francis Group
Boca Raton London New York

CRC Press is an imprint of the
Taylor & Francis Group, an **informa** business

Originally published in journal form as **Philosophical Transactions of The Royal Society**, series A, Volume 352, Number 1699, pp 197–385. © 1995 The Royal Society.

First published 1996 by Gordon and Breach Science

Published 2019 by CRC Press
Taylor & Francis Group
6000 Broken Sound Parkway NW, Suite 300
Boca Raton, FL 33487-2742

First issued in paperback 2020

ISBN-13: 978-0-367-57946-3 (pbk)
ISBN-13: 978-90-5699-020-6 (hbk)

Visit the Taylor & Francis Web site at
http://www.taylorandfrancis.com

and the CRC Press Web site at
http://www.crcpress.com

British Library Cataloguing in Publication Data

The Arctic and environmental change
  1. Ecology — Arctic regions 2. Man — Influence on nature —
Arctic regions 3. Greenhouse effect, Atmospheric — Arctic
regions 4. Arctic regions — Environmental conditions
I. Wadhams, P. II. Dowdeswell, Julian A., 1957–
III. Schofield, A. N.
333.7'0998

ISBN 9056990209

CONTENTS

Contents

Preface

On October 12–13 1994 the Royal Society held a Discussion Meeting on the Arctic and Environmental Change. The fourteen papers which were presented form the basis of this book. They constitute a wide-ranging review of the process of climate change as it affects the Arctic, and give a broad insight into the transformation of the Arctic which we can expect during the next century on account of anthropogenic warming.

The special importance of the Arctic to the global warming problem is made clear in the first paper by Cattle and Crossley, which shows results from the latest General Circulation Model (GCM) produced by the Hadley Centre for Climate Prediction and Research. This is one of a small number of GCMs produced by leading climate modelling centres throughout the world, with all the models being in general agreement about regional trends in climate change. The consensus view is that within 70 years the Arctic is predicted to warm by at least 4°C on an annual average and in places by more than 8°C, as compared to values of 0–4°C elsewhere in the world and even a slight cooling in parts of the Southern Ocean. The warming will be greater in the winter, and this large Arctic enhancement of global warming persists even when the moderating effect of sulphate aerosols is taken into account. One mechanism contributing to this enhancement is ice-albedo feedback, whereby the retreat of seasonal snow on land, and sea ice in the ocean, reduces average albedo and so generates a positive feedback loop. Since this meeting, the Hadley model runs have been further reported by Mitchell *et al.* (1995), and the consensus view of the anticipated extent of global warming (and its regional variation) is being reported in the updated Scientific Assessment of the Intergovernmental Panel on Climate Change.

If these predictions are correct, it is in the Arctic that we can expect to observe global warming at its most powerful. A series of chapters deals with atmospheric effects related to climate change. Serreze *et al.* discuss the variability of atmospheric circulation and water vapour fluxes in the Arctic, while McIntyre and Pyle discuss in theoretical and observational terms how the stratospheric polar vortex operates and how it may be perturbed to allow a significant ozone loss in the Arctic as well as the Antarctic. Stanhill offers a cautionary note with an analysis of solar irradiance and how this is affected by atmospheric pollution so as to lessen any warming effect. Northern Hemisphere ozone loss in late winter appears to have become more serious since the time of the meeting; as this Preface goes to press, there are reports that the thickness of the ozone layer over Britain sank from a seasonal (early March) average of 365 Dobson Units to record low levels of 195 DU at Lerwick on 5 March 1996 and 206 DU at Camborne on 3 March (Pearce, 1996). This was associated with exceptionally cold temperatures in the Arctic stratosphere.

Moving on to impacts, Callaghan describes the effects of climatic warming on terrestrial ecology and Gradinger on marine ecology. Perhaps the most significant effects will be on the ocean and cryosphere. Rudels shows how ocean thermohaline structure is affected by the reduced amount of cooling and sea ice production in winter. Wadhams discusses observed and expected changes to Arctic sea ice extent and thickness, and Dowdeswell the impacts on Arctic glaciers, where a wholesale retreat would make a significant contribution to sea level rise. Wingham presents evidence from satellites that the Greenland Ice Sheet is itself experiencing fluctuations in volume. On land, an important change will be the melting and retreat of permafrost; Williams discusses the likely effects.

The importance of the oceanic changes discussed in the last section of Rudels' paper is underlined by new discoveries made by the Canadian–US Arctic Ocean transect carried out in 1994 by the icebreakers USCGC *Polar Sea* and CCGS *Louis S. St. Laurent*. Preliminary results were announced by Aagaard (1995) at the Wadati Conference on Global Change and the Polar Climate held at Tsukuba, Japan, on 7–10 November 1995, and a paper is in press (Carmack *et al.*, 1996). The cruise discovered large changes in the thermal structure of the Arctic Ocean over a wide depth range, apparently due to advection (i.e. a greater heat input from the West Spitsbergen Current). There is a significant overall warming of the Atlantic layer, concentrated high in the water column at about 200 m depth. This is accompanied by wholesale displacement of halocline waters of Pacific origin by colder nutrient-poor waters from the Eurasian Basin. The temperature anomaly appears to have entered the Arctic Ocean from the Norwegian Sea in the late 1980s. The decline in Greenland Sea convection discussed by Rudels is also the subject of continuing study by the EC-sponsored European Subpolar Ocean Programme, with results from this and the earlier Greenland Sea Project being reviewed in a 1995 conference in Hamburg (AOSB, 1995). It is clear from both these strands of research that major rearrangements of the thermohaline structure of the northern high-latitude ocean have been occurring in recent years, and this is bound to be a most active area of further research.

The meeting ended with two papers on past climates, as revealed both by Greenland ice cores and by sediment cores in the polar North Atlantic. Dowdeswell and White demonstrate the very rapid rates of past climate change in ice-core records. Thiede discusses the timing of the inception of Northern Hemisphere ice sheets, and the palaeoceanography of the Norwegian-Greenland Sea. Core analysis is an area in which very rapid advances are taking place, and, as with ocean research, since the Royal Society meeting significant new findings have continued to emerge, in this case from studies of the 3,000 m GRIP and GISP-2 ice cores from central Greenland. Temperature profiles from these two boreholes have been used to provide calibrations for palaeo-temperatures inferred from the oxygen isotope records at each site (Cuffey *et al.*, 1995; Johnsen *et al.*, 1995). These analyses, achieved using different datasets and techniques, have indicated that glacial-interglacial temperature shifts at the summit of the Greenland Ice Sheet were, at about 20°C, almost twice as large as had previously been thought. This has major implications for the nature of atmospheric and oceanic reorganisation at the end of the last glacial.

On the other hand, detailed physical inspection of the GRIP and GISP-2 cores below about 2,800 m (about 110,000 years ago) has shown that significant structural disturbance, presumably a result of ice flow, is present in both records (Alley *et al.*, 1995). This means that the rapid, large magnitude fluctuations in isotopic values, and thus in inferred palaeo-temperature, during the Eemian interglacial, are artefacts of core disturbance. The GRIP ice-core record cannot, therefore, be used as evidence for an unstable last interglacial climate, and the apparent contrasts with environmental stability during the present or Holocene interglacial are not supported. Other evidence, however, e.g. from measurements of magnetic susceptibility, pollen and organic carbon in lake deposits (Thouveny *et al.*, 1994), does still point towards the possibility that rapid climate events did occur in the Eemian, while there is much evidence in the ice cores for rapid events in more recent glacial periods.

New Ocean Drilling Program (ODP) cores, collected in Autumn 1995 as part of the North Atlantic Arctic Gateways ODP Leg 162, confirm significant contrasts in the timing of the initiation of major glaciation on the Norwegian and Greenland sides of the Polar North Atlantic. Significant iceberg rafting, implying glaciation to sea level, began about 7 million years ago, in the late Miocene, off central East Greenland, confirming earlier ODP findings to the south (Larsen *et al.*, 1994). By contrast, large-scale glaciation with iceberg production began at 2.5 to 3 million years ago on the eastern margin of the Polar North Atlantic (Jansen and Sjøholm, 1991). This variability in the timing of inception of large-scale glaciation in the North Atlantic region demonstrates clearly the spatial variability in palaeoclimate that has affected this part of the Arctic.

After the Royal Society meeting an informal discussion meeting was held on the following day under the auspices of the CIBA Foundation, attended by many of the contributors to the Royal Society meeting. The topic was "rapid climatic change", and the result which stimulated much discussion was the dataset from Greenland ice cores which shows a relatively stable climate since the recovery from the last full glacial period, but rapid short-term fluctuations before that, sometimes including excursions of several degrees in air temperature occurring within a few decades. As mentioned above, there is now a question as to whether these fluctuations are real. However, the possible causes of pre-anthropogenic fluctuations, and the possibility of the reappearance of an unstable climate, independently of Man's activities, were problems which the delegates regarded as of central importance.

When scientists have explained recent geological history in popular terms they have up to now expressed uncertainty as to whether the late Cenozoic ice age is yet over, or whether the present time is simply an interglacial interval. It is clear that both the North American and Eurasian ice sheets went through a number of cycles of formation, expansion and recession during the Pleistocene, with glacial episodes separated by warmer inter-glacial intervals. The recent data from Greenland ice cores also show that, whether or not unexpected rapid climate changes occurred during the Eemian interglaciation (about 115,000 years ago), the rises in global temperature and in sea level since the beginning of the Holocene (about 10,000 years ago) have been smooth and continuous, more so than during any glacial period in the record.

The retreat of the ice, and the smooth and steady warming, led to the development of grasslands at high latitudes all round the world, an environment in which the northern nations of mankind came into existence. It was the relative stability of the Holocene climate that allowed human society to develop rapidly in the past 10,000 years. If any human activities or unknown natural processes can provoke unexpected rapid climate changes, it is important for our entire future as a species to be able to understand and predict them, and if possible avoid them. Yet it is clear that much more needs to be known before rapid climate changes can be understood. The Arctic, considered to be the home of the most rapid anthropogenically induced changes, is a natural laboratory where climate changes and their impacts can be monitored and studied more readily than elsewhere in the world. Continued research in the Arctic is therefore vital for understanding our climatic future.

References

Aagaard, K. 1995 The recent warming of the Arctic Ocean. In *Program and Preprints, Wadati Conference on Global Change and the Polar Climate, 7–10 November, Tsukuba Science City, Japan.* Geophysical Inst., University of Alaska, Fairbanks, 25–27.

Alley, R. B., Gow, A. J., Johnsen, S. J., Kipfstuhl, J., Meese, D. A. & Thorsteinsson, T. 1995 Comparison of deep ice cores. *Nature*, **373**, 393–394.

AOSB 1995 *"Nordic Seas", Hamburg, March 7–9 1995, Extended Abstracts.* Published by Arctic Ocean Sciences Board and Sonderforschungsbereich "Processes Relevant to Climate", 234pp.

Carmack, E. C., Aagaard, K., Swift, J., Perkin, R. G., McLaughlin, F. A., Macdonald, R. W., Jones, E. P., Smith, J. N., Ellis, K. & Kilius, L. 1996 Large thermohaline changes in the Arctic Ocean. *Nature* (in press).

Cuffey, K. M., Clow, G. D., Alley, R. B., Stuiver, M., Waddington, E. D. & Saltus, R. W. 1995 Large Arctic temperature change at the Wisconsin-Holocene glacial transition. *Science*, **270**, 455–458.

Jansen, E. & Sjøholm, J. 1991 Reconstruction of glaciation over the past 6 Myr from ice-borne deposits in the Norwegian Sea. *Nature*, **349**, 600–603.

Johnsen, S. J., Dahl-Jensen, D., Dansgaard, W. & Gundestrup, N. S. 1995 Greenland temperatures derived from GRIP borehole temperature and ice core isotope profiles. *Tellus*, **47B**, 624–629.

Larsen, H. C., Saunders, A. D., Clift, P. D., Beget, J., Wei, W., Spezzaferri, S. & ODP Leg 152 Scientific Party. 1994 Seven million years of glaciation in Greenland. *Science*, **264**, 952–955.

Mitchell, J. F. B., Johns, T. C., Gregory, J. M. & Tett, S. F. B. 1995 Climate response to increasing levels of greenhouse gas and sulphate aerosols. *Nature*, **376**, 501–504.

Pearce, F. 1996 Big freeze digs a deeper hole in ozone layer. *New Scientist*, **149(2021)**, 7.

Thoveny, N., de Beaulieu, J-L., Bonifay, E., Creer, K. M., Guiot, J., Icole, M., Johnsen, S., Jouzel, J., Reille, M., Williams, T. & Williamson, D. 1994 Climate variations in Europe over the past 140 kyr deduced from rock magnetism. *Nature*, **371**, 503–506.

P. WADHAMS

J.A. DOWDESWELL

A.N. SCHOFIELD

Modelling Arctic climate change

BY H. CATTLE AND J. CROSSLEY

Hadley Centre for Climate Prediction and Research,
Meteorological Office, Bracknell, Berkshire, RG12 2SY

Climate prediction requires the use of coupled models of the atmosphere–deep ocean–sea ice and land surface. This paper outlines the formulation of processes relevant to the simulation and prediction of climate change in the Arctic of one such model, that of the Hadley Centre for Climate Prediction and Research at the Meteorological Office. Comparison of the simulation of a number of features of the Arctic climate is made against observations and predictions of future climate change resulting from increased concentrations of greenhouse gases from recent runs of the model are discussed.

1. Introduction

Prediction of future climate change due to increased concentrations of greenhouse gases resulting from the burning of fossil fuels and other human activities requires the use of coupled models of the climate system. Such models consist of a number of component physical models (of the atmosphere, land surface, oceans and sea ice) which are interactively 'coupled' by exchange of information across the interfaces between them. Thus, for example, the fluxes (of heat, freshwater and momentum) which drive the ocean are calculated within the atmospheric model and passed across to the ocean model. The ocean model in turn calculates new values of sea surface temperature which are passed back to the atmospheric model in the next phase of its integration. This, then, provides an interactive coupling between the atmosphere and ocean model components.

The aim in the construction of such models is to represent the key processes and feedbacks important for climate prediction and for studies of climate variability. Previously they have shown the Arctic to be a region of high climate sensitivity to increased concentrations of greenhouse gases (see Houghton *et al.* 1990, 1992). In this paper we describe some aspects of the Arctic simulation of one such model, that of the Hadley Centre for Climate Prediction and Research at the Meteorological Office, and outline its current projections of future climate change over the region. However, it must be borne in mind that confidence in the ability of climate models to represent regional climate change is as yet rather low (Houghton *et al.* 1990, 1992). Nevertheless, the sensitivity of the Arctic to increased concentrations of greenhouse gases is a consistent feature of global climate model simulations. This is seen both in sensitivity studies to instantaneous doubling of greenhouse gases using atmospheric models coupled to simple 'slab' ocean models (see Manabe & Stouffer 1980; Washington & Meehl 1986; Wilson & Mitchell 1987; see also Houghton *et al.* 1990, 1992) and in model runs to study the 'transient' response to gradually increasing concentrations of greenhouse gases (e.g. Manabe *et al.* 1991, 1992; Cubash *et al.* 1992; Murphy 1995; Murphy & Mitchell 1995). Such runs require atmospheric models to be coupled to dynamical models of the ocean, allowing ocean processes to be represented throughout its entire depth. For a recent review of the Arctic sensitivity of models to increased greenhouse gases, see Rowntree (1993).

2. Model description

The model used in the experiments described here is a version of the Meteorological Office unified model for numerical weather prediction and climate studies (Cullen 1993). In climate mode, the atmospheric model is run coupled to ocean, sea ice and land surface models on a 2.5° × 3.75° latitude–longitude grid. The atmospheric model has 19 levels in the vertical and the ocean model 20. The atmospheric model incorporates the interactive radiation scheme of Ingram (1990), the prognostic cloud scheme of Smith (1990) (which includes explicit representation of cloud liquid water), a penetrative convection parametrization based on Gregory & Rowntree (1990), but with the addition of a representation of the effects of convective downdraughts and a gravity wave drag scheme based on the work of Palmer *et al.* (1986). The ocean model is based on that of Cox (1984) with the addition of an upper-ocean mixed-layer scheme based on Kraus & Turner (1967), the shear induced mixing parametrization of Pacanowski & Philander (1981) and the isopycnal mixing scheme of Redi (1982). The land surface scheme is based on that of Warrilow *et al.* (1986).

A key feature of high-latitude climate is its snow and ice cover. In the model, snow falling on the land surface accumulates there when the surface temperature is below 0 °C. Snow depth is allowed to change interactively over the model's land surface through the processes of accumulation, sublimation/deposition and melting. Over land, a snow albedo formulation is used which depends on snow depth, the snow-free albedo of the surface, a deep snow albedo for temperature and vegetation type and, near the melting point, on temperature. Surface temperature, T_*, changes as a function of time, t, according to the prognostic equation:

$$C_* \partial T / \partial t = R_N + H + LE + G, \tag{1}$$

where C_* is the effective surface thermal capacity, R_N is the net solar plus longwave surface radiative flux, H is the surface sensible heat flux, E is the surface evaporative flux, L the latent heat of vaporization (or, if the suface is snow covered, of sublimation) and G is the soil heat flux. If snow is present, then snow melt is allowed to occur when the surface temperature is predicted by (1) to rise above 0 °C. In that case, the surface temperature is reset to the melting point and the excess energy is used to melt snow.

Snow accumulation and melting on the surface of sea ice is also represented in the model. Sea ice forms within the ocean model when the surface temperature of the surface layer over a given model timestep falls below the freezing point of sea water (−1.8 °C) or as a result of advection out from the ice edge. In that case, the layer temperature is reset to the freezing point and a sufficient covering of sea ice (of initial mean thickness 0.5 m) is allowed to form to ensure that conservation of heat is satisfied. The model thus allows fractional sea-ice coverage. The formulation of fractional ice cover is based on that of Hibler (1979), which assumes the ice covered area to have a uniform ice thickness distribution of mean thickness h_I. The atmospheric model calculates surface radiative and turbulent heat fluxes separately for ice and open water (leads) over grid squares which include sea ice.

A simple ice thickness advection scheme is used in the model following Bryan *et al.* (1975), in which the rate of change of mean ice thickness over a grid square changes according to:

$$\partial h_{\text{I}}/\partial t = \nabla \cdot (\delta_h \boldsymbol{v} h_{\text{I}}) + A_H \nabla^2 h_{\text{I}} + \text{thermodynamic changes}, \qquad (2)$$

where $\boldsymbol{v}$ is the ocean current vector in the surface layer of the model ocean; A_H is the ocean thermal diffusivity and $\delta_h = 1$ for $h < 4$ m, 0 for $h > 4$ m. Though simple, introduction of this formulation produces a marked improvement in the model's representation of the seasonal variation of sea ice in the southern hemisphere in particular (cf. Cattle *et al.* 1993), and removes the need for flux correction (see below) of the sea ice itself, though these are still applied to ocean beneath the ice.

The sea ice thermodynamics formulation used in the model follows the zero-layer model of Semtner (1976) in which the ice–snow layer is treated as a single slab. The surface temperature of the ice changes according to the same heat balance equation as (1), in which G now becomes the heat conduction through the ice, H_{I}, given by the equation:

$$H_{\text{I}} = k_{\text{S}}(T_{\text{S}} - T_{\text{F}})/[h_{\text{S}} + (h_{\text{I}} k_{\text{S}}/k_{\text{I}})], \qquad (3)$$

where k_{S} and k_{I} are the thermal diffusivities of snow and ice; h_{S} and h_{I} the snow and ice layer thicknesses and T_{S} and T_{F} the ice/snow surface and ice bottom temperatures respectively. T_{F} is assumed to be at the freezing point of seawater (taken as -1.8 °C), as is the surface temperature of leads.

For surface temperatures below -10 °C, the snow/ice surface albedo is taken to have a constant value of 0.8. Above this temperature, ice albedo decreases linearly to a value of 0.5 at the melting point (0 °C) to allow for the lowering of albedo caused by the presence of melt ponds on the ice surface. The albedo of leads is assumed to be a constant 0.06.

Any net heat flux entering (or leaving) the leads is partitioned between ice melt (or ice formation) and warming (or cooling) of the upper layer of the ocean. The partitioning between ice melt/formation and ocean warming/cooling is chosen to be directly proportional to the ice area. Ocean temperatures less than -1.8 °C also result in ice formation. Ocean surface layer temperatures higher than -1.8 °C result in a bottom heat flux from the ocean to the ice given by:

$$H_{\text{o}} = \rho c k(T_1 - T_{\text{F}})/0.5\Delta z_1,$$

where ρ is the density and c the specific heat capacity of seawater, k is an 'eddy diffusivity', taken as 2.5×10^{-3} m^2 s^{-1} and T_1 is the temperature of the uppermost layer of the model ocean, of depth Δz_1.

At the ice surface, snow melt (or, if no snow is present, ice melt) occurs in the model under the same conditions as for snow on the land surface. Any meltwater formed is passed to, and freshens, the surface layer of the ocean model. Alternatively, brine release associated with ice formation leads to reduced freshening. More generally over the model oceans, salinity can change depending on the sign of the precipitation minus evaporation difference. Surface freshening (important for the Arctic salinity balance) can also occur as a result of runoff from the model's land surface which is formulated following the scheme described by Taylor & Bunton (1993).

Because of the relatively long timescales on which they respond to change, the

major land ice sheets of Greenland and Antarctica are treated non-interactively in
the model with the ice mass held fixed and specified through the model's topography
field. A constant surface albedo of 0.8 is used over land ice.

3. Model performance over the Arctic

In this section, we illustrate the ability of the coupled system to represent some
of the major features of Arctic climate. The fields shown here are from a run of the
model with constant greenhouse gas concentrations used as a control against which to
compare scenario runs with increased greenhouse gas concentrations (Hadley Centre
1995). The runs followed a long spin-up integration of the coupled system in which
sea surface temperatures and salinities were relaxed back to their seasonally varying
climatological observed values. Thus, for example, in specifying the net heat flux into
the ocean, this quantity was modified by the addition of a term:

$$-\lambda(T_m - T_c)$$

where T_m and T_c are the model's instantaneous and the observed climatological sea
surface temperatures (ssts) respectively and λ is a relaxation coefficient (taken in
the runs described here to have a value of 165 W m^{-2}K^{-1}). Towards the end of the
spinup integration, these terms were averaged to form seasonally and geographically
varying flux correction fields which were then applied during the control and scenario
runs. Such an artifact is commonly used in the present generation of climate models
(see Cubasch *et al.* 1992; Manabe *et al.* 1991; Murphy 1995) and is necessary to
remove the drifts which would otherwise occur in the model climate.

Figure 1 shows the surface mean sea level pressure field over the Arctic for January
and July as observed and as simulated by the model. Proper simulation of the surface pressure field is important since the associated wind field provides the primary
dynamical forcing on sea ice. In winter (figure 1a), the observed surface pressure field
is dominated over the Atlantic by the Icelandic low which extends northeastwards
over the Greenland Sea into the eastern Arctic. A ridge of high pressure extending
across the Arctic to the north of the Bering Strait links the winter high pressure systems over the American and Eurasian continental areas; to the south of the Bering
Strait lies the Aleutian low. This pattern begins to evolve during March and April
towards the summer pattern of figure 1b in which a weak ridge extends across the
Arctic linking the high pressure systems extending northeastwards from the Pacific
and Azores highs. The return to the winter pattern begins to take place in September. The model gives a good representation of the seasonal evolution of the surface
pressure patterns, illustrated by figures 1c and d, which show the simulated patterns
for January and July.

For January, the model shows the Icelandic trough in winter to be associated
with the zone of maximum precipitation and cloud, with rather lower cloud amounts
and low precipitation values lying along the high pressure ridge over the central
Arctic (figure 2c and d). These features are borne out in observational fields given
by Legates & Willmott (1990) (figure 2a and b) for precipitation and, for example,
in the cloudiness fields shown by Orvig (1970). The seasonal evolution of modelled
total cloud amount for the area north of 60° N is compared with surface-based
observed values in figure 3. The model shows perhaps too much cloud overall, with a
reduced seasonal cycle, though the climatology is uncertain, especially in winter. The
modelled seasonal evolution of precipitation over sea ice polewards of 75° N (figure 4)

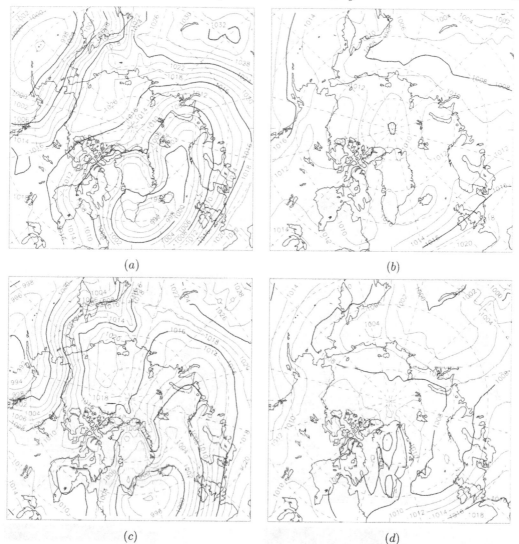

Figure 1. Mean sea level pressure over the Arctic as observed for (a) January and (b) July and modelled for (c) January and (d) July. Observed fields are derived from UK Meteorological Office operational analyses.

shows a marked seasonal variation with area averaged precipitation a minimum in spring and a maximum in autumn. Overall amounts over the year given by the model are somewhat higher overall than the typical annual mean figure of 0.5 mm day^{-1} (see the dataset of Legates & Willmott (1990)), though, again, the climatology over the central Arctic basin is uncertain.

The pattern of surface temperature shown by the model for January is shown in figure 5a. Observed fields (see Orvig 1970) show surface air temperatures in January to be coldest (down to −50 °C or more) over the Greenland ice cap, and the Siberian land mass, where lowest mean temperatures are below −40 °C. Over the Arctic basin itself, the coldest air is found on the northern side of the Canadian Archipelago with average temperatures down to −34 °C. A tongue of warmer air penetrates the Arctic via the northeastwards extension of the Icelandic trough and, to a lesser extent, in the

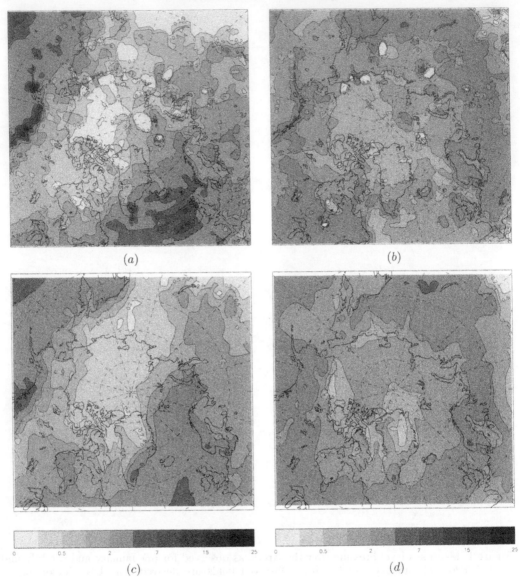

Figure 2. As figure 1, but for precipitation. Observed fields are derived from the dataset of
Legates & Willmott (1990).

region of the Bering Strait. Overall, these features are well captured by the model. In
summer (figure 5b), surface temperatures over the pack ice are constrained to lie near
freezing by the melting snow and ice surface. Comparison of the seasonal evolution
of ice surface temperature polewards of 75° N (not shown) with the observations
of Untersteiner (1960) (see Semtner 1976) over perennial sea ice shows the model
to reproduce the seasonal evolution of surface temperature fairly faithfully, though,
compared to the data, values are a little too high in spring and early winter.

Overall, the model underestimates the total sea ice extent for the Arctic and
overestimates the amplitude of the seasonal variation of this quantity. Though the
peak wintertime extents almost match the observed (figure 6), the spring meltback
is too rapid with the result that for much of the year extents (and concentrations,

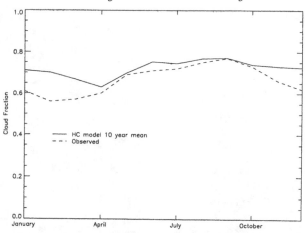

Figure 3. Seasonal variation of mean cloudiness polewards of 60° N as modelled (—) and derived from surface-based observations (- - -).

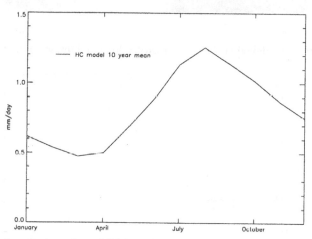

Figure 4. Seasonal variation of modelled total precipitation over the Arctic north of 75° N (mm day^{-1}).

particularly in summer) are markedly low. Ice thicknesses also tend to be too low in both winter and summer, with the model failing to achieve the thick ice (of order 5 m) observed to occur against the north coast of Greenland and the Canadian Archipelago. The model shows maximum ice thicknesses in the central Arctic of just over 3 m in March and some 2 m in September.

Over much of the Arctic basin, the seasonal cycle of freezing and melting of sea ice, coupled with inflow from the Arctic river systems combine to produce a surface layer of cold and relatively fresh water some tens of metres deep (figure 7a). Beneath this, to a depth of 100–150 m, lies the Arctic halocline in which salinity increases markedly with depth, but temperatures remain close to the freezing point. Below the halocline, salinity increases more slowly with depth, but temperatures increase quickly as the underlying layer of water of Atlantic origin is reached. Low resolution ocean models of the type used here employ large coefficients of viscosity for numerical stability reasons (Bryan *et al.* 1975) with the result that current strengths tend to be poorly simulated (higher resolution is precluded by available computing resources for

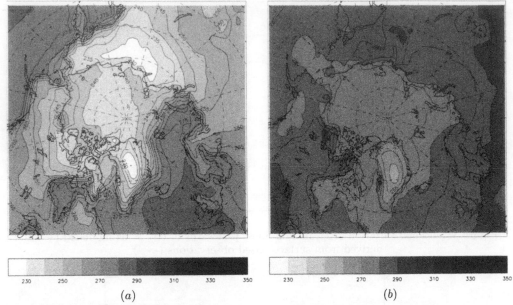

(a)   (b)

Figure 5. Modelled surface temperature for (a) January and (b) July.

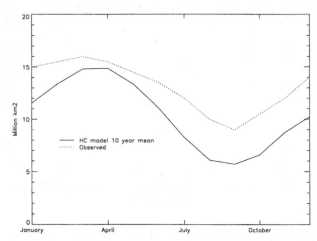

Figure 6. Seasonal variation of modelled (solid line) and observed (dotted line) total northern hemisphere ice extent. Observed values are digitized from Parkinson *et al.* (1987)

century-timescale runs). Nevertheless, the model clearly shows the inflow of Atlantic water into the Arctic basin (centred in this case at about 500 m). It also simulates (figure 7b) the effects of freshwater stabilization of the upper waters of the Arctic, but with a halocline which lacks the sharpness of the observed halocline and without reaching the correct degree of warmth in the nose of the temperature profile.

4. Simulation of Arctic climate change

Figure 8 shows predicted surface temperature change from an integration of the coupled model in which atmospheric concentrations of greenhouse gases (based on carbon dioxide, CO_2, as a surrogate) are increased at their observed rate from 1860

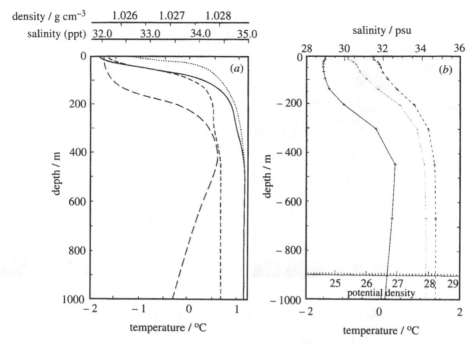

Figure 7. (a) Composite vertical distributions of temperature from Aagaard & Coachman (1975) (– –), salinity ( ) and density (- - -) for the Arctic Ocean and salinity for the southern Eurasian basin (···). (b) Model profiles for the Eurasian Arctic: —, temperature; - - -, salinity; ···, potential density.

to the present day and then increased further at 1% per annum (compound) into the future. For further details of the runs, see Hadley Centre (1995). The changes shown are for the decade 2020–2030, that of doubling of CO_2 concentrations in the model. Consistent with results from other models, the largest global temperature increases occur over the Arctic in winter (taken as the period December–February). Maximum temperature rises are associated with the marginal ice zone in the Atlantic sector and the regions of the shelf seas (figure 8a). Changes over the ice mass of Greenland lie between 2–4 °C. Introduction of a simple parametrization of the effects of sulphate aerosols reduces the magnitude of the warming, but changes the overall pattern very little (J. F. B. Mitchell, personal communication).

In summer (figure 8b), the temperature change over the Arctic is shown by the model to be small, since, as already noted, surface temperatures are constrained to the melting point of sea ice. Figure 9 shows the sea ice itself is reduced in thickness, by over 1 m in both summer and winter, with maximum changes coincident with the thickest ice in the control run.

Precipitation changes are shown in figure 10. The model predicts winter precipitation to be increased slightly over the central Arctic basin with higher local increases over the surrounding land masses. A general drying is shown over the region of the Greenland–Iceland–Norwegian (GIN) Sea, but with increases over the Atlantic-sector marginal ice zone. In summer, there is a tendency towards reduced precipitation around the periphery of the Arctic Ocean, with regions of both slight increase and decrease in the central basin. The precipitation changes show some sensitivity to introduction of aerosol effects in the model formulation. These result in an increase

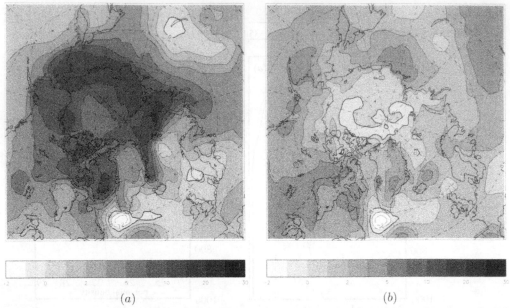

(a)   (b)

Figure 8. Temperature change over the Arctic for the decade of doubling of carbon dioxide from a run of the Hadley Centre model with transiently increasing greenhouse gases: (a) winter (December–February) and (b) summer (June–August).

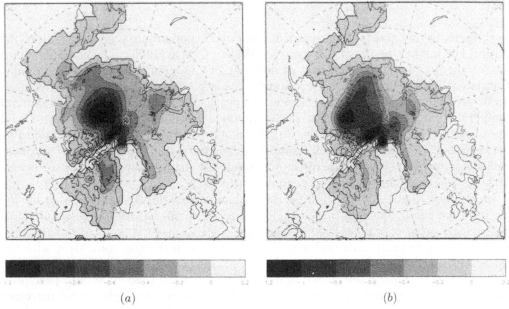

(a)   (b)

Figure 9. As figure 8, but for change in sea ice thickness.

in the areas of reduced precipitation in over the land areas around the periphery of the Arctic in winter, and more generally in summer.

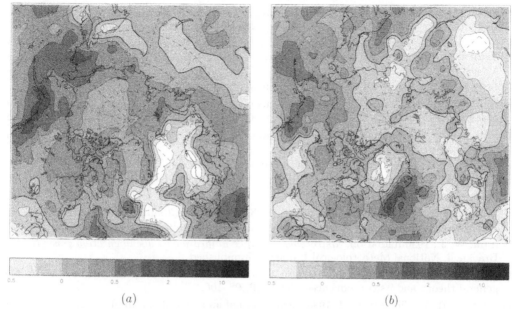

(a) (b)

Figure 10. As figure 8, but for change in precipitation.

5. Concluding remarks

Consistent with other model simulations, the Hadley Centre model shows the Arctic to be a region of high sensitivity to increased greenhouse gas concentrations in winter. The largest temperature changes over the Arctic Ocean are to be found in the region of the marginal ice zone and the shelf seas, and there is a generally reduced ice thickness over the central Arctic, with largest values (of over 1 m) coincident with the thickest ice in the control simulation. There is a general small increase in precipitation over the Arctic in winter and regionally variable small increases and decreases in summer.

It must, of course, be born in mind that, as already noted, confidence in predictions on a regional basis using global coupled models is still, as yet, relatively low. Comparison of results of Arctic warming in previous transient response experiments by different centres certainly reveals large differences between models (Rowntree 1993). Uncertainties in both an Arctic and wider, global context are associated, in particular, with sulphate aerosols, clouds, ocean resolution, representation of sea ice processes and representation of runoff into the Arctic. Nevertheless, the comparison of model and observed climatology shown here is encouraging and helps to give credence to the large-scale characteristics of the predictions.

Many people within the Hadley Centre have been involved in the development of the coupled model described here and their contribution is gratefully acknowledged. Particular thanks are due to Dr T. C. Johns and Dr S. F. B. Tett and who ran the experiments described in this paper.

References

Aagaard, K. & Coachman, L. K. 1975 Toward an ice-free Arctic Ocean. *EOS* **56**, 484–486.

Bryan, K., Manabe, S. & Pacanowski, R. C. 1975 A global ocean–atmosphere climate model. II. The oceanic circulation. *J. phys. Oceanogr.*, **5**, 30–46.

Cox, M. D. 1984 A primitive equation, 3-dimensional model of the ocean. *GFDL Ocean Group tech. Rep.* No. 1. Princeton, NJ: Geophysical Fluid Dynamics Laboratory.

Cubasch, U., Hasselmann, K., Hock, H., Maier-Reimer, E., Mikolajewicz, U., Santer, B. D. & Sausen, R. 1992 Time dependent greenhouse warming computations with a coupled ocean–atmosphere model. *Climate Dyn.* **8**, 55–69.

Cullen, M. P. J. 1993 The unified forecast/climate model. *Meteorol. Mag.* **122**, 81–94.

Gregory, D. & Rowntree, P. R. 1990 A mass flux convection scheme with representation of cloud ensemble characteristics and stability dependent closure. *Mon. Weather Rev.* **118**, 1483–1506.

Hadley Centre 1995 *Modelling Climate Change 1860–2050.* Bracknell: Hadley Centre, Meteorological Office.

Hibler, W. D. III 1979 A dynamic-thermodynamic sea ice model. *J. phys. Oceanogr.* **9**, 817–846.

Houghton, J. T., Jenkins, G. J. & Ephraums, J. J. (eds) 1990 *Climate change: the IPCC scientific assessment.* Cambridge University Press.

Houghton, J. T. Callander, B. A. & Varney, S. K. (eds) 1992 *Climate change 1992: the supplementary report to the IPCC scientific assessment.* Cambridge University Press.

Ingram, W. J., 1990 Radiation. *Meteorological Office unified model documentation paper No. 23.* Bracknell: National Meteorological Library. (Unpublished.)

Kraus, E. B. & Turner, J. S. 1967 A one dimensional model of the seasonal thermocline. II. The general theory and its consequences. *Tellus* **19**, 98–106.

Legates, D. R. & Willmott, C. J. 1990 Mean seasonal and spatial variability in gauge-corrected global precipitation. *Int. J. Climatol.* **10**, 111–127.

Manabe, S., Stouffer, R. J., Spelman, M. J. & Bryan, K. 1991 Transient responses of a coupled ocean–atmosphere model to gradual changes of atmospheric CO_2. Part 1. Annual mean response. *J. Climate* **4**, 785–818.

Manabe, S., Spelman, M. J. & Stouffer, R. J. 1992 Transient responses of a coupled ocean–atmosphere model to gradual changes of atmospheric CO_2. Part 2. Seasonal response. *J. Climate* **5**, 105–126.

Manabe, S. & Stoufer, R. J. 1980 Sensitivity of a global climate model to an increase in CO_2 concentration in the atmosphere. *J. geophys. Res.* **85**, 5529–5554.

Mitchell, J. F. B. & Murphy, J. M. 1995 Transient response of the Hadley Centre coupled ocean–atmosphere model to increasing carbon dioxide. Part 2. Spatial and temporal structure and response. *J. Climate* **8**, 57–80.

Murphy, J. M. 1995 Transient response of the Hadley Centre coupled ocean–atmosphere model to increasing carbon dioxide. Part 1. Control climate and flux adjustment. *J. Climate* **8**, 36–56.

Orwig, S. (ed.) 1970 *World survey of climatology*, vol. 14. *Climates of the polar regions.* Amsterdam: Elsevier.

Pacanowski, R. C. & Philander, S. G. H. 1981 Parametrization of vertical mixing in numerical models of tropical oceans. *J. phys. Oceanogr.* **11**, 1143–1451.

Palmer, T. N., Schutts, G. J. & Swinbank, R. 1986 Alleviation of systematic bias in general circulation and numerical weather prediction models through orographic gravity wave drag parametrization. *Q. Jl R. Met. Soc. Lond.* **112**, 1001–1039.

Parkinson, C. L., Comiso, J. C., Cavalieri, D. J., Gloersen, P. & Campbell, W. J. 1987 *Arctic sea ice, 1973–1976,* NASA SP-489.

Redi, M. H. 1982 Oceanic isopycnal mixing by coordinate rotation. *J. phys. Oceanogr.* **12**, 1154–1158.

Rowntree, P. R. 1993 Global and regional patterns of climate change: recent predictions for the Arctic. *Clim. Res. tech. Note* No. 43. Bracknell: Meteorological Office.

Semtner, A. J. Jr. 1986 A model for the thermodynamic growth of sea ice in numerical investigations of climate. *J. Phys Oceanogr.* **6**, 379–389.

Smith, R. N. B. 1990 A scheme for predicting layer clouds and their water content in a general circulation model. *Q. Jl R. Met. Soc. Lond.* **116**, 435–460.

Taylor, N. K. & Bunton, C. 1993 River runoff in the new UKMO coupled model. *Research Activities in Atmospheric and Oceanic Modelling* (ed. G. J. Boer), Report No. 18, WMO/TD-No. 553. Geneva: World Meteorological Organisation.

Warrilow, D. A., Sangster, A. B. & Slingo, A. 1986 Modelling of land surface processes and their influence on European climate. *Dyn. Clim. tech. Note* No. 38. Bracknell: Meteorological Office.

Washington, W. M. & Meehl, G. A. 1986 General circulation model CO_2 sensitivity experiments: sea ice albedo parametrizations and globally averaged surface air temperature. *Clim. Change* **8**, 231–241.

Wilson, C. A. & Mitchell, J. F. B. 1987 A doubled CO_2 sensitivity experiment with a GCM including a simple ocean. *J. geophys. Res.* **92**, 13315–13343.

Discussion

D. J. DREWRY (*NERC, Swindon, UK*). Model validation and interpretation of model results rely fundamentally upon accurate observations. In the Arctic there are only very limited and spatially patchy datasets: time series of data are short in duration. How do these limitations influence and affect the Arctic and regional predictions?

H. CATTLE. The models do simulate the gross features of Arctic climate. However, while the models represent the important feedbacks in the climate system, their details are imperfectly represented and this will affect the predictions in a way which is difficult to quantify. An increased observational database for the Arctic is certainly needed both for model verification and for better representation in models of Arctic physical processes. This is a primary aim of, for example, the World Climate Research Programme 10-year Arctic Climate System Study.

Variability in atmospheric circulation and moisture flux over the Arctic

By Mark C. Serreze[1], Roger G. Barry[1],
Mark C. Rehder[1] and John E. Walsh[2]

[1] Cooperative Institute for Research in Environmental Sciences,
Division of Cryospheric and Polar Processes,
University of Colorado, Boulder CO 80309, USA
[2] Department of Atmospheric Sciences, University of Illinois,
Urbana–Champaign, Urbana IL 61801, USA

Mean characteristics and variability in the spatio-temporal distribution of Arctic water vapour and vapour fluxes are examined using several different rawinsonde-derived databases. Precipitable water averaged over the polar cap, 70–90° N, peaks in July at 14.0 mm. Large poleward fluxes near the prime meridian reflect transport associated with north Atlantic cyclones and, for most months, a local maximum in available water vapour. The mean vapour flux convergence averaged for the polar cap peaks in September. There is a mean annual excess of precipitation minus evaporation $(P - E)$ of 163 mm, with a 78 mm range between extreme years. High $P - E$ is favoured by a meridional circulation accompanied by a more dominant North Atlantic cyclone track. No trend in annual $P - E$ is apparent over the 1974–1991 period.

1. Introduction

Uncertainties in the distribution of Arctic water vapour can cause errors in satellite-derived estimates of surface albedo (Rossow *et al.* 1989) and ice surface temperatures (Key & Haefliger 1992). Changes in water vapour flux convergence and precipitation minus evaporation $(P - E)$ may alter river runoff into the Arctic Ocean, with consequent changes in the density structure of the upper ocean, possibly influencing sea ice production (Cattle 1985), and through ice advection, deep-water formation in the peripheral seas (Mysak *et al.* 1990). Although the Arctic response to anthropogenic greenhouse warming will likely be amplified due to the temperature–albedo feedback, an attendant increase in atmospheric moisture, through its added greenhouse effect, is expected to further enhance warming (Raval & Ramanathan 1989). Changes in the Arctic atmospheric moisture budget are also likely to be manifested by altered cloud cover, further impacting on the surface radiation budget. The characteristics of Arctic water vapour have been addressed as a component of global studies (e.g. Peixoto & Oort 1983; Gaffen *et al.* 1991; Peixoto & Oort 1992), for specific regions in the Arctic (e.g. Barry 1967), for means at 70° N (Serreze *et al.* 1994a) as well as for the entire Arctic Basin (Burova 1983; Burova & Voskresenskii 1976; Drozdov *et al.* 1976; Serreze *et al.* 1995a), but the need exists for a synthesis of the basic features of its variability. Here, we draw from some of our recent studies and summarize

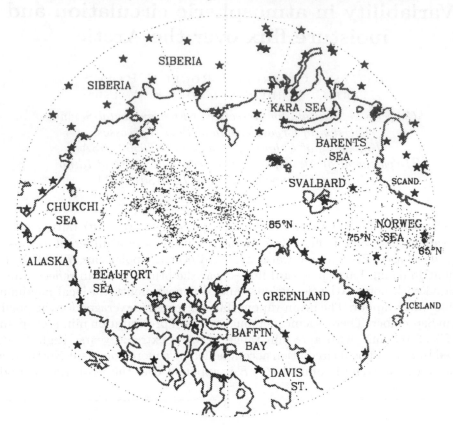

Figure 1. The distribution of fixed stations in the HARA during 1988 (stars) and drifting ice station and ship soundings (dots, plotted at every 10th sounding).

the characteristics of Arctic water vapour and its transports, using data from three rawinsonde archives.

2. Data and quality control

(a) Rawinsonde archives

The first dataset, the Historical Arctic Rawinsonde Archive (HARA) (Kahl *et al.* 1992), contains soundings taken one to four times per day for fixed stations north of 65° N. Records of at least 30 years are available for about 60 stations, typically extending through 1991. The second archive comprises soundings taken from 1–2 times daily from the Russian 'North Pole' (NP) series of drifting ice stations in the Arctic Ocean. The data coverage extends from 1954 through 1990, and includes records from NP 3–4, 6–17 19, 21–22, 26 and 28. While often only a single station was in operation at any one time, there are many periods with overlapping records from multiple stations. Some stations contain records of several years or more. Approximately 17 000 soundings are available. The third archive comprises approximately 16 500 soundings made from ships. These soundings, available from 1976–1991 at either 0000 or 1200 UTC, were primarily taken over the Norwegian and Barents seas (figure 1).

All soundings provide measurements of temperature, humidity and winds at both fixed mandatory-reporting pressure levels (e.g. surface, 850 mbar, 700 mbar) and significant pressure levels (intermediate levels based on vertical-change criteria specified for the reporting of upper-air sounding data). Both data types are used here. The HARA and ship soundings typically extend up to at least 300 mbar. By contrast, the majority of the NP soundings extend only to about 700 mbar. We use all of the NP and ship data. As the number of soundings per station and number of significant levels reported were fewer prior to 1974 for many Eurasian stations in the HARA, we use HARA data for the 1974–1991 period only.

(b) Quality control and processing

All data were subject to extensive quality control procedures. These consist of a series of vertical consistency and limits checks discussed by Serreze *et al.* (1994a, c). Erroneous or questionable data values were discarded, and then refilled through vertical interpolation. The Canadian HARA station data often had missing surface winds. Consequently, missing surface winds in the HARA were obtained from a climatological relationship between surface and 850 mbar wind direction and wind speed, compiled from a five-year subset of all HARA soundings (Serreze *et al.* 1994a). Soundings over the Arctic Ocean rarely had missing surface winds; if one of these sounding had missing surface winds, it was simply discarded. Three summary datasets were then compiled. In the first dataset (D1), the 0000 and 1200 UTC HARA soundings for 1974–1991 were processed to obtain monthly station means of precipitable water (from the vertical integral of specific humidity) and vertically integrated vapour fluxes for five layers (surface–850 mbar, 850–700 mbar, 700–500 mbar, 500–400 mbar and 400–300 mbar). The station means were then passed into a Cressman (1959) interpolation procedure to provide data at 70° N at every 10° of longitude, allowing for analyses of the climatological transport 'pathways' of water vapour into and out of the Arctic, as well as estimates of climatological vapour flux convergence and precipitation minus evaporation ($P - E$) over the north polar cap (70–90° N). $P - E$ is estimated as (Alestalo 1983)

$$P - E = -\nabla \cdot \boldsymbol{F}_\mathrm{m} - \partial Q/\partial t, \tag{1}$$

where $-\nabla \cdot \boldsymbol{F}_\mathrm{m}$ is the horizontal convergence of the vertically integrated (surface to 300 mbar) meridional vapour flux ($\boldsymbol{F}_\mathrm{m}$) across 70° N, and $\partial Q/\partial t$ is the monthly change in precipitable water (Q) based on the means of all stations north of 70° N. The $\partial Q/\partial t$ term is typically 5–20% of the magnitude of the flux convergence, being largest in the transitional months (Serreze *et al.* 1994a). The flux convergence is

$$-\nabla \cdot \boldsymbol{F}_\mathrm{m} = \frac{1}{A} \oint \boldsymbol{F}_\mathrm{m} \, \mathrm{d}C, \tag{2}$$

where $\mathrm{d}C$ is the length along the 70° N circle (totalling 1.37×10^4 km) and A is the area enclosed (1.54×10^7 km^2). Interpolated fluxes, the flux convergence and $P - E$ were obtained for each month of each year, with the results for individual years then averaged. To be included in the analysis, a station for a given month and year had to have valid data for at least 75% of all possible days. For the second database (D2), monthly HARA station means not passing the 75% data-availability threshold were discarded, with new means then found via Cressman interpolation from surrounding stations. This resulted in a complete time series for all stations. Meridional fluxes, the flux convergence and $P - E$ over the north polar cap for each month and year were then determined by passing into the Cressman interpolation

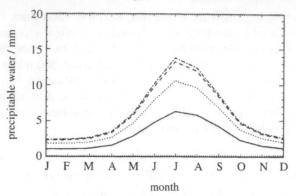

Figure 2. Mean monthly precipitable water (mm) averaged over the 70–90° N region for the surface–850 mbar (—), surface–700 mbar (···), surface–500 mbar (- - -) and surface–300 mbar (— · — · —) layers.

to 70° N only those station means requiring initial interpolation from surrounding stations for less than 30 of the possible 216 months in the 18-year data record. Long-term mean values are nearly identical to those derived from the D1 set. However, by minimizing the effects of changes in the station network, the technique provides for an internally consistent assessment of interannual variability in flux convergence and $P - E$. We have also compiled a gridded database (D3). Briefly, all ship and drifting station soundings were passed into the Cressman procedure to obtain climatological monthly means of precipitable water and vertically integrated vapour fluxes at NMC grid locations (Octagonal Grid Format) over the ocean. These grid-point means were then passed into a second interpolation that included long-term monthly means at the HARA stations, resulting in climatological monthly fields for all NMC grid locations north of 65° N. Details are provided by Serreze *et al.* (1994*b, c*). Elliot & Gaffen (1991) and Garand *et al.* (1992) discuss uncertainties in water vapour analyses at low temperatures and relative humidities and differences between countries in instrument types and reporting practices. Nevertheless, all of our analyses use the sounding data essentially as given. Since instrument changes and reporting practices have undergone frequent changes, and are often poorly documented, accounting for them fully would be impractical. Furthermore, as demonstrated in a series of sensitivity tests by Serreze *et al.* (1993*c*), these inhomogeneities may at worst result in potential errors of 5% in vertically integrated moisture variables (used here). This reflects the fact that problems in rawinsonde humidity data tend occur at the higher levels and lower temperatures at which water vapour is negligible.

3. Results

(*a*) *Precipitable water*

Figure 2 shows the seasonal cycle in precipitable water for 70–90° N from the surface–850 mbar, surface–700 mbar, surface–500 mbar and surface–300 mbar, based on the D3 dataset. Assuming that water vapour is negligible above 300 mbar, total precipitable water (surface–300 mbar) ranges by a factor of over five from about 2.5 mm in January to 14.0 mm in July, when surface and tropospheric temperatures are also highest. These values agree closely with those given by Peixoto & Oort (1992) based on an earlier dataset (1963–1973) with less vertical resolution. For all months,

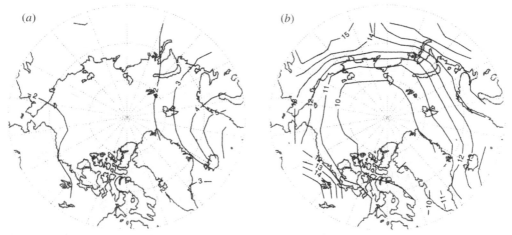

Figure 3. Mean precipitable water (mm) over the Arctic Basin in the surface–700 mbar layer for: (*a*) January, (*b*) July. Contours over the high (greater than 3000 m) Greenland ice cap have been omitted.

approximately 80% of precipitable water is found within the surface–700 mbar layer, and about 95% in the surface–500 mbar layer.

Figure 3 shows the spatial field of surface–700 mbar precipitable water for January and July from the D3 dataset. While gridded fields are available for the surface–300 mbar, fewer data points represent the ocean grid-point means above 700 mbar. For January (figure 3*a*), precipitable water increases from the central Arctic Ocean, Eurasia and Canada (less than 2 mm) towards the Atlantic side of the Arctic, peaking between Iceland and Scandinavia at 4–6 mm. This reflects the higher tropospheric temperatures in this region (Gorshkov 1983), allowing the atmosphere to hold more moisture. Qualitatively, the same pattern of precipitable water occurs from October through April. By contrast, the July pattern (figure 3*b*), qualitatively representative of May–September, is much more zonal, reflecting the more zonal temperature distribution.

(*b*) *Moisture flux*

Corresponding fields of the vertically integrated meridional vapour flux (surface–700 mbar) are provided in figure 4. For January (figure 4*a*), peak poleward transports are found between Iceland and Scandinavia (greater than 30 kg m^{-1} s^{-1}). This represents the combined effects of the regional maximum in available moisture and moisture transports associated with cyclonic activity along the primary North Atlantic storm track (Serreze *et al.* 1995*a*) (figure 5). A maximum in the poleward meridional flux in this region is observed for all months. The July fluxes are much larger (figure 4*b*); although the atmospheric circulation is weaker than in winter (Serreze *et al.* 1993), large fluxes are possible due to the greater availability of moisture. The pronounced area of negative (equatorward) fluxes over northern Canada in July (weakly present in January) relates to persistent equatorward winds on the eastern limb of the western North American longwave ridge. For all months, fluxes tend to peak in the lower troposphere, typically at about 850 mbar. This represents the 'trade-off' level between the effects of specific humidity decreasing with height, and winds increasing with height (Serreze *et al.* 1994*a, c*).

Figure 6 shows the longitudinal distribution of the vertically integrated (surface to

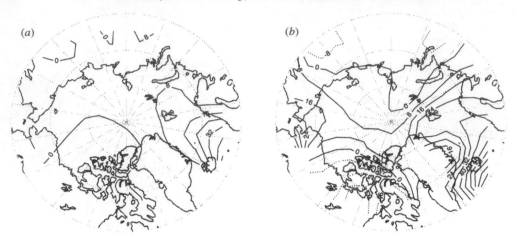

Figure 4. Mean vertically integrated meridional vapour flux (kg m^{-1} s^{-1}) over the Arctic Basin for the surface–700 mbar layer for (a) January and (b) July. Contours over the high (greater than 3000 m) Greenland ice cap have been omitted.

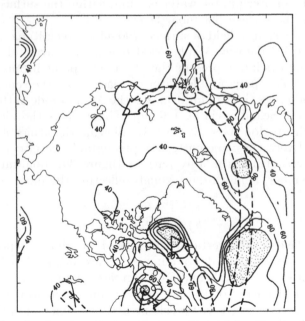

Figure 5. Counts of sea level cyclone centers at NMC grid point locations and generalized cyclone tracks for winter. Stipples denote areas with more than 100 cyclones (unpublished, based on the Serreze et al. (1993) cyclone detection algorithm).

300 mbar) meridional flux for January, July and September at 70° N, based on the D1 dataset. Each month shows peak poleward transports near the prime meridian and Baffin Bay, consistent with frequent cyclonic activity near these areas. A subsidiary peak is found near Alaska, best expressed in July and September. September also shows a fairly strong peak near 90° E. All months also show equatorward fluxes over the Canadian Arctic Archipelago.

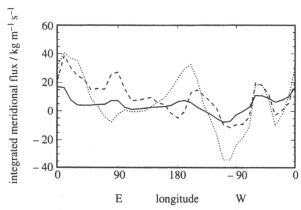

Figure 6. Longitudinal distribution of the mean (1974–1991) vertically integrated meridional water vapour flux (surface–300 mbar, in kg m^{-1} s^{-1}) at 70° N for January (—), July ($\cdots$) and September (- - -).

(c) Moisture flux convergence

The mean flux convergence into the polar cap is positive for all months. The monthly value, in terms of water depth averaged over the 70–90° N region, is about 10 mm for December through February, compared with 14 mm for July, and 22 mm in September, which represents the maximum for the year. As precipitable water is strongly decreasing in September (figure 2), $P - E$ also obtains its annual maximum of 26 mm in this month. The mean annual total $P - E$ of 163 mm is about 35% higher than that reported by Peixoto & Oort (1992), but compares closely with a value of 169 cm for the Arctic Ocean cited by Burova (1983). She also estimates that the contribution of precipitation from moisture evaporated and condensed which falls out within the region to be about 2/3 of the local evaporation (i.e. 120 mm) with a further contribution of 30 mm from condensation on the surface. Substituting into our $P - E$ value a climatological estimate of E for the 70–90° N region from the Korzun (1976) atlas (based on coastal and drifting station data), yields an areal-average annual precipitation of 266 mm, which compares favourably with the value of 293 cm for the same region from re-digitized precipitation values in the Gorshkov (1983) atlas. The reason why the flux convergence peaks in September, two months after peak precipitable water in July (figure 2), is apparent in figure 6. In July, the large poleward fluxes near the prime meridian, Alaska and over Baffin Bay are compensated by large moisture outflows over the Canadian Arctic Archipelago. By contrast, although the fluxes are generally more modest in September, there is less compensation between regions of inflow and outflow. Note in particular the poleward fluxes found from the Baffin Bay region eastward to about 140° E.

(d) Time series of flux coverage and P − E

Figure 7 provides the time series of the seasonal flux convergence expressed as liquid water averaged over the 70–90° N region for the period 1974–1991, based on the D2 dataset. For each year, the individual bar segments from bottom to top represent, respectively, the flux convergences for winter (January–March), spring (April–June), summer (July–September) and autumn (October–December). As over each annual cycle, $\partial Q/\partial t$ sums to zero, the sum of the seasonal flux convergences is annual $P - E$. The seasonal definitions used to compile figure 7 hence facilitate

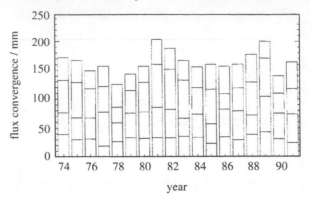

Figure 7. Time series of seasonal vapour flux convergence and annual $P - E$ (mm) averaged over the 70–90° N region, 1974–1991 (see text).

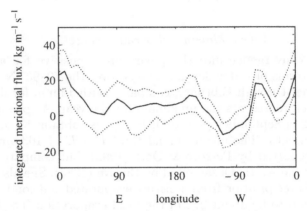

Figure 8. Longitudinal distribution of the mean annual vertically integrated meridional water vapour flux (1974–1991) and range between extreme years (surface–300 mbar. in kg m^{-1} s^{-1})

evaluation of seasonal contributions of the flux convergence to annual $P - E$ for standard calendar years. Annual $P - E$ ranges by 78 mm from a low of 125 mm in 1978 to a high of 203 mm in 1981. Although there is no apparent trend in $P - E$, there is some suggestion of a multi-year cycle, with higher values in the early to mid 1970s and early and late 1980s, which needs to be tested with a longer record. The mean seasonal flux convergences are 32, 44, 51 and 38 mm, respectively for winter, spring, summer and autumn.

(e) Causes of variability

Figure 8 shows the mean annual meridional flux by longitude at 70° N and the maximum and minimum at each longitude over the 1974–1991 period, based on the D2 dataset. The range between extreme years is highest (typically greater than 20 kg m^{-1} s^{-1}) from the prime meridian to 130° E and least around 60° W. A similar pattern is found for the seasonal flux values. This leads one to suspect that inter-annual variability in the seasonal flux convergence and annual $P - E$ are strongly controlled by flux variations over the Eurasian (0–130° E) sector. Composite analyses by Serreze *et al.* (1995*b*) of 500 mbar height, based on the three highest and lowest flux convergences for each season (not shown) reveal that large flux convergences are favoured by a more meridional 'winter type' circulation, characterized by strong eastern North American and East Asian troughs. In turn, analyses of the distribution of sea level cyclone centers reveals a more pronounced North Atlantic cyclone track, with more cyclonic activity near the prime meridian and extending into the Kara Sea (winter, spring and autumn) promoting large vapour transports, or in summer, local increases over the Arctic Ocean accompanied by sharp reductions in cyclonic activity over Eurasia and Canada. Normally, summer cyclonic activity is frequent in these areas (Serreze *et al.* 1993). The low composites display essentially the opposite pattern. Variability in the strength of dominance of the North Atlantic cyclone track appears largely responsible for the large range in the meridional flux in the Eurasian sector, and hence variations in the flux convergence. Fluxes of liquid water, not considered here, would also be influenced by cyclonic activity.

4. Summary and conclusions

Using several rawinsonde archives, we have analyzed aspects of Arctic water vapour and its transports. For the 70–90° N region, surface–300 mbar precipitable water ranges from about 2.5 mm in January to 14.0 mm in July, with about 80% of the total column water vapour found in the surface to 700 mbar layer. During winter, peak values occur over the Atlantic side of the Arctic, reflecting the higher tropospheric temperatures, while during summer, a more zonal distribution is observed.

The meridional vapour flux exhibits large seasonal and spatial variability. Peak poleward transports found near the prime meridian manifest the greater availability of water vapour and moisture advection along the North Atlantic cyclone track. In turn, equatorward transports over the Canadian Arctic Archipelago reflect persistent northerly winds. These patterns are particularly apparent in the longitudinal distribution of the vertically integrated meridional flux across 70° N. The vapour flux convergence is positive for all months, peaking in September. The mean annual inflow into the Arctic corresponds to an excess of precipitation over evaporation of 163 mm. For the 1974–1991 period, $P - E$ ranges from a low of 125 mm in 1978, to a high of 203 mm in 1981, but no trends are apparent. Large seasonal flux convergences are favoured by a meridional 'winter type' circulation, characterized by a stronger or more dominant North Atlantic cyclone track.

This study was supported by NSF grants DPP-9214838 and DPP-9113673. The NSIDC staff is thanked for computer support.

References

Alestalo, M. 1983 The atmospheric water vapour budget over Europe. In *Variations in the global water budget* (ed. A. Street-Perrott, M. Beran & R. Ratcliffe), pp. 67–79. Dordrecht, The Netherlands: D. Riedel Publishing Co.

Barry, R. G. 1967 Variations in the content and flux of water vapour over north-eastern North America during two winter seasons. *Q. J. R. Met. Soc. Lond.* **93**, 535–543.

Burova, L. P. 1983 *Vlagooborot v atmosfere arktike* (Moisture exchange in the Arctic atmosphere), 128 pp. Leningrad: Gidrometeoizdat.

Burova, L. P. & Voskresenskii, A. I. 1976 Soderzhanie i perenos vlagi v atmosfere nad severnoi poliarnoi oblast'iu (Atmospheric moisture content and transport over the north polar area). *Leningrad, Arkticheskii i antarkticheskii nauchno-issledovatel'skii institut, Trudy* **323**, 25–39.

Cattle, H. 1985 Diverting Soviet rivers: some possible repercussions for the Arctic Ocean. *Polar Record* **22**, 485–498.

Cressman, G. P. 1959 An operational objective analysis system. *Mon. Wea. Rev.* **87**, 367–374.

Drozdov, O. A., Sorochan, O. G., Voskresenskii, A. I., Burova, L. P. & Kryshko, O. V. 1976 Kharakteriskiki vlagooborota v atmosfere nad skonami basseina severnogo ledovitogo okeana (Characteristics of the atmospheric water budget over Arctic Ocean drainage basins). *Leningrad, Arkticheskii i antarkticheskii nauchno-issledovatel'skii institut, Trudy* **327**, 15–34.

Elliot, W. P. & Gaffen, D. J. 1991 On the utility of radiosonde humidity archived for climate studies. *Bull. Am. meteor. Soc.* **72**, 1507–1520.

Gaffen, D. J., Barnett, T. P. & Elliot, W. P. 1991 Space and time scales of global tropospheric moisture. *J. Climate* **4**, 989–1008.

Garand, L., Grassotti, C., Halle J. & Klein, G. 1992 On differences in radiosonde humidity-reporting practices and their implications for numerical weather prediction and remote sensing. *Bull. Am. meteor. Soc.* **73**, 1417–1423.

Gorshkov, S. G. 1983 *World Ocean Atlas, Volume 3: Arctic Ocean* (in Russian), 184 pp. plus appendices. Oxford: Pergamon Press.

Kahl, J. D., Serreze, M. C., Shiotani, S., Skony, S. M. & Schnell, R. C. 1992 *In situ* meteorological sounding archives for Arctic studies. *Bull. Am. meteor. Soc.* **73**, 1824–1830.

Key, J. & Haefliger, M. 1992 Arctic ice surface temperature retrieval from AVHRR thermal channels. *J. Geophys. Res.* **97**, 5885–5893.

Korzun, V. I. 1976 *Atlas of world water balance,* 122 pp. Leningrad: Gidrometeiozdat.

Mysak, L. A., Manak, D. K. & Marsden, R. F. 1990 Sea ice anomalies observed in the Greenland and Labrador seas during 1901–1984 and their relation to an interdecadal Arctic climate cycle. *Climate Dynamics* **5**, 111–133.

Peixoto, J. P. & Oort, A. H. 1983 The atmospheric branch of the hydrological cycle and climate. In *Variations in the global water budget* (ed. A. Street-Perrott, M. Beran & R. Ratcliffe), pp. 5–65. Dordrecht, The Netherlands: D. Reidel Publishing Co.

Peixoto, J. P. & Oort, A. H. 1992 *Physics of climate,* 520 pp. New York: American Institute of Physics.

Raval, A. & Ramanathan, V. 1989 Observational determination of the greenhouse effect. *Nature, Lond.* **342**, 758–761.

Rossow, W. B., Brest, C. L. & Gardner, L. C. 1989 Global, seasonal surface variations from satellite radiance measurements. *J. Climate* **2**, 214–247.

Serreze, M. C., Box, J. E., Barry, R. G. & Walsh, J. E. 1993 Characteristics of Arctic synoptic activity, 1952–1989. *Meteor. atmos. Phys.* **51**, 147–164.

Serreze, M. C., Barry, R. G. & Walsh, J. E. 1994*a* Atmospheric water vapour characteristics at 70° N. *J. Climate* **8**, 719–731.

Serreze, M. C., Rehder, M. C., Barry, R. G. & Kahl, J. D. 1994*b* A climatological database of Arctic water vapor characteristics. *Polar Geog. Geol.* **18**, 63–75.

Serreze, M. C., Rehder, M. C., Barry, R. G., Kahl, J. D. & Zaitseva, N. A. 1995*a* The distribution and transport of atmospheric water vapor over the Arctic Ocean. *Int. J. Climatol.* (In the press.)

Serreze, M. C., Rehder, M. C., Barry, R. G., Walsh, J. E. & Robinson, D. A. 1995*b* Variations in aerologically-derived Arctic precipitation and snowfall. *Ann. Glaciol.* (In the press.)

Discussion

D. Drewry (*NERC, Polaris House, Swindon, UK*). Are the radiosonde data sufficient to examine regional scale questions such as trends in precipitation and consequential mass balance over the Greenland Ice Sheet?

R. G. Barry. In selected regions, the radiosonde network is adequate to compute the net moisture balance over a polygonal area. However, this approach is not recommended for small regions due to local anomalies in the wind components. Rasmusson

(1968) suggested that areas 2.0×10^6 km^2 or larger are required. Adequate temporal sampling of all synoptic weather events is equally critical.

The mass balance of the Greenland area (2.1×10^6 km^2) has in fact recently been examined using rawinsonde data by Robasky & Bromwich (1994). A decreasing trend of accumulation during 1963 to 1988 is indicated.

Walsh *et al.* (1994) similarly examine the moisture balance for the Mackenzie drainage basin (2.2×10^6 km^2 area).

Additional references

Rasmusson, E. M. 1968 Atmospheric water vapor transport and the water balance of North America. II. Large-scale water balance investigations. *Mon. Weather Rev.* **96**, 720–734.

Robasky, F. M. & Bromwich D. H. 1994 Greenland precipitation estimates from the atmospheric moisture budget. *Geophys. Res. Let.* **21**, 2495–2498.

Walsh, J. E., Zhou, X., Portis, D. & Serreze, M. C. 1994 Atmospheric contributions to hydrologic variations in the Arctic. *Atmos.-Ocean* **32**, 733–755.

The stratospheric polar vortex and sub-vortex: fluid dynamics and midlatitude ozone loss

By M. E. McIntyre

Centre for Atmospheric Science at the Department of Applied Mathematics and Theoretical Physics, Silver Street, Cambridge CB3 9EW, UK

It has been suggested on the basis of certain chemical observations that the winter-time stratospheric polar vortex might act as a chemical processor, or flow reactor, through which large amounts of air – of the order of one vortex mass per month or three vortex masses per winter – flow downwards and then outwards to middle latitudes in the lower stratosphere. If such a flow were to exist, then most of the air involved would become chemically 'activated', or primed for ozone destruction, while passing through the low temperatures of the vortex where fast heterogeneous reactions can take place on polar-stratospheric-cloud particles. There could be serious implications for our understanding of ozone-hole chemistry and for midlatitude ozone loss, both in the Northern and in the Southern Hemisphere. This paper will briefly assess current fluid-dynamical thinking about flow through the vortex. It is concluded that the vortex typically cannot sustain an average throughput much greater than about a sixth of a vortex mass per month, or half a vortex mass per winter, unless a large and hitherto unknown mean circumferential force acts persistently on the vortex in an eastward or 'spin-up' sense, prograde with the Earth's rotation. By contrast, the 'sub-vortex' below pressure-altitudes of about 70 hPa (more precisely, on isentropic surfaces below potential temperatures of about 400 K) is capable of relatively large mass throughput depending, however, on tropospheric weather beneath, concerning which observational data are sparse.

1. Introduction

There is controversy over the causes of the observed midlatitude decline in stratospheric ozone (e.g. Pyle, this volume). The main questions are (*a*) whether the midlatitude decline represents polar ozone loss that is spreading, in some sense, to middle latitudes, (*b*) whether such spreading involves further ozone loss in middle latitudes, or mere dilution by ozone-depleted polar air, and (*c*) how far the total midlatitude decline in stratospheric ozone might proceed in future. These problems concern the Northern as well as the Southern Hemisphere. Although the chemical ozone-loss mechanisms are generally strongest in the Southern Hemisphere, the fluid-dynamical transport mechanisms are generally, as it happens, strongest in the Northern Hemisphere.

What is our most secure piece of knowledge about these problems? It is that the most conspicuous ozone loss seen so far, which takes place in the springtime Antarctic stratosphere, is mostly caused by man-made halocarbons. These stable chemicals are observed to be well mixed in the troposphere, with Northern and Southern Hemispheric mixing ratios generally within about 10% of each other. This is possible because of the halocarbons' exceptional chemical stability, and low solubility in water. The halocarbons are carried up into the tropical stratosphere, and taken to high stratospheric altitudes in a slow but inexorable large-scale upwelling motion; they are then photolysed by the Sun's hard ultraviolet radiation to which they become exposed above about 20 km altitude. This global-scale pattern of tropospheric mixing and large-scale, tropical-stratospheric upwelling is part of the reason why Northern Hemispheric pollution can cause Southern Hemispheric ozone loss.

The mechanism of the large-scale tropical upwelling is well understood, being describable to first approximation as a global-scale 'gyroscopic pumping'. The extratropical stratosphere and mesosphere, up to altitudes of about 80 km, act on the tropical stratosphere like a gigantic, seasonally and interannually variable suction pump, whose action depends on the Earth's rotation together with certain wavelike and turbulent eddy motions. Further explanation and references are given in the Appendix. The effect is to pull air gently but persistently upward and poleward out of the tropical troposphere and lower stratosphere, then push it back down toward the extratropical troposphere, most of it through the winter stratosphere via complicated, chaotic pathways. Some of the poleward and downward moving air, carrying photolysed halocarbon fragments, gets into the polar lower stratosphere where, in the case of the Antarctic, the photolysed fragments participate in the formation of a conspicuous 'ozone hole' via reactions during winter and spring whose final stages depend on the springtime return of sunlight. Meanwhile, more tropospheric air is being pulled up into the tropical stratosphere, importing more halocarbons to be photolysed in their turn.

Typical large-scale upwelling velocities in the tropical lower stratosphere (altitudes 15–20 km) are seasonally variable roughly from 0.2 mm s^{-1} in northern summer to 0.4 mm s^{-1} in northern winter, or roughly 6–13 kilometres per year with the largest values confined mainly to the northern winter. Such velocities are of course not directly measurable but there are now several independent ways of estimating them, including new and relatively precise tracer information (Mote *et al.* 1995; Holton *et al.* 1995). The gyroscopic pumping rate sets the e-folding timescale, of the order 100 years, for removal of the halocarbons from the troposphere, because rates of land and ocean uptake of halocarbons are at least a decimal order of magnitude slower (Junge 1976).

The gyroscopic pumping mechanism has often been called 'wave driving' and the resulting global-scale circulation the 'wave-driven circulation', because of the nature of the wavelike and turbulent eddy motions involved (see the Appendix), giving rise for instance to the so-called 'Rossby-wave surf zone' in middle latitudes. These same wavelike and turbulent eddy motions give rise to another remarkable and chemically important phenomenon, the approximate chemical isolation of the winter stratospheric polar vortex, now widely believed to be important for ozone-hole chemistry. There is strong evidence, both observational and theoretical – for historical background see McIntyre (1989) and for recent observational evidence see, e.g., among many other references, Dahlberg & Bowman (1994), Lahoz *et al.* (1995), Manney *et al.* (1994*b*), Waugh *et al.* (1993), Norton & Chipperfield (1995) – there is strong evidence that the edge of the stratospheric vortex acts as a flexible 'eddy-transport barrier', bounding the surf zone and inhibiting the eddy transport of lagrangian tracers along the stratification or isentropic surfaces and across the edge of the vortex.

Fast eddy transport along isentropic surfaces would be expected to take place

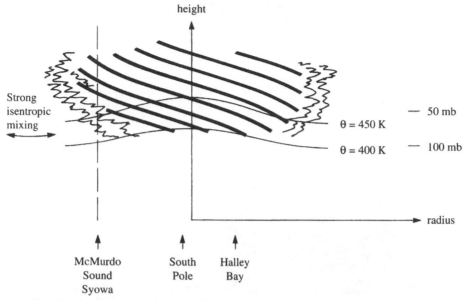

Figure 1. Rough schematic cross-section through the lower-stratospheric part of a polar vortex, updating figure 5 of McIntyre (1989) where it was pointed out (*a*) that the jagged edge is to be expected from the general properties of layerwise-two-dimensional turbulence in the surrounding 'Rossby-wave surf zone', and (*b*) that such a structure could explain some or all of the layering seen in balloon soundings from McMurdo Sound, Antarctica. McMurdo Sound is often near the vortex edge in the Antarctic winter.

freely on the basis of the usual intuitions about quasi-horizontal or layerwise-two-dimensional turbulence in strongly stratified fluids. Such fast eddy transport does indeed take place, but only in the surf zone and similar regions. The spatial inhomogeneity vitiates standard turbulence-theoretic assumptions, but it is an almost inevitable consequence of a relevant fluid-dynamical theorem, the conservation of Rossby–Ertel potential vorticity (PV). Such inhomogeneity is practically certain to occur in real flows of the kind in question, for simple but robust theoretical reasons (McIntyre 1994). A typical example was shown at the Discussion Meeting using a videotape made from a high-resolution numerical model stratospheric simulation (Norton 1994), and available as a VHS PAL or NTSC cassette on request from Dr W. A. Norton, Dept. of Atmospheric, Oceanic and Planetary Physics, Oxford (wan@atm.ox.ac.uk).

This videotape demonstrates all the effects in question, in particular the dramatic way in which the eddy-transport barrier marking the edge of the polar vortex can strongly inhibit eddy tracer transport, even in the presence of a violently distorted vortex and strong layerwise-two-dimensional turbulence in the midlatitude Rossby-wave surf zone. The barrier effect depends on two things, first the 'Rossby-wave quasi-elasticity' associated with isentropic gradients of PV, which gradients tend to be concentrated in the vortex edge, and second the horizontal shear near the edge of the vortex (Juckes & McIntyre 1987).

The likely fine structure of the vortex edge is shown schematically in figure 1. The effect of the surrounding 'surf zone' is to erode the edge and steepen the isentropic gradients of PV and chemical tracers. There is usually a multiple-edge structure in these gradients, here suggested only schematically. This has been demonstrated

M. E. McIntyre

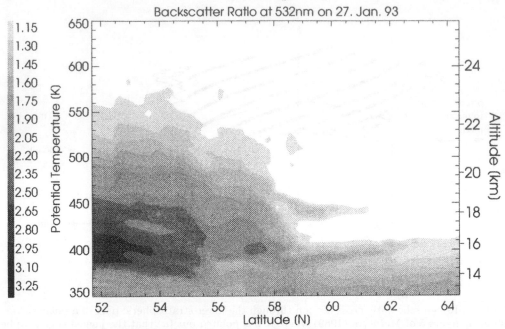

Figure 2. Airborne lidar cross-section through the side and bottom of the Arctic polar vortex: compare bottom part of the left-hand edge in figure 1. The lower part of the vortex. and the eddy-transport barrier constituting its edge, is made visible by the presence of the stratospheric aerosol layer or Junge layer, extending to its usual midlatitude altitudes of about 550 K. but largely excluded from the vortex interior (in right half of figure). Note that potential temperature is used as the vertical coordinate; kilometre altitudes shown on the right are only approximate. Grayscale values give the so-called backscatter ratio – a measure of aerosol mixing ratio. or mass of aerosol per unit mass of air – at a visible wavelength strongly scattered by typical stratospheric aerosol particles. From unpublished research extending the work of Dameris *et al.* (1995): courtesy of Dr Martin Wirth.

especially clearly by the recent work of Waugh *et al.* (1993) using high-resolution lagrangian advection techniques and meteorological estimates of the wind field. Because of the combined effects of surf-zone turbulence and vertical shear. the edge is likely to show a 'screwthread' structure. left-handed in the Antarctic and right-handed in the Arctic. with slopes of the order of Prandtl's ratio (Coriolis parameter over stratification buoyancy frequency. of the order 10^{-2}). The screwthread is probably not as regular as is shown schematically here for the sake of visual clarity. Manney *et al.* (1994*a*) show, in a relatively low-resolution numerical model experiment. the initial stages of formation of such a structure. Figure 2 shows a lidar slice through the Arctic vortex and surrounding aerosol, suggesting a somewhat coarser structure than in figure 1. whose bottom left corner may be compared to the lowest part of the white region in figure 2.

2. The controversy over polar-vortex isolation: the 'flowing processor' hypothesis

Despite the foregoing considerations. the effectiveness of the transport-inhibition mechanism in the real stratosphere has recently been called into question (Tuck *et al.* 1993 and references therein). Tuck *et al.* and others have tentatively hypothesized that, even in the Antarctic, there is a large flow of air through the real polar vortex, downwards and then outwards to middle latitudes in the lower stratosphere. Such a flow would take large amounts of air and halocarbon photolysis fragments through the vortex, where they would be chemically 'activated' by fast heterogeneous reactions within the colder parts of the vortex, then exported to sunlit middle latitudes in the lower stratosphere, the densest part of the ozone layer, with possibly serious consequences in the form of ozone depletion over populated areas. The hypothesized picture has been summarized in the phrase 'the vortex as a flowing processor', or 'flow reactor'.

The flow through the vortex is hypothesized to persist throughout winter and early spring, to be a major contributor to the observed midlatitude ozone decline, both in the Northern and in the Southern Hemisphere, and to explain other observed phenomena as well – particularly the dryness of air in the extratropical Southern Hemispheric lower stratosphere that has been observed on some occasions. This dryness, in other words, is hypothesized to be due mainly to the dehydration of air transported through the very cold Antarctic vortex (Tuck *et al.* 1993 & references), rather than through the coldest parts of the tropical lower stratosphere, or through the polar 'sub-vortex' region below the 400 K isentropic surface, roughly corresponding to pressure-altitudes below 70 hPa.

Tuck *et al.* (1993 & references) state that the vortex is 'flushed several times' during a single winter. This presumably means that the flushing time, τ_F say, defined by

$$\tau_F = M/\dot{M}, \tag{1}$$

is of the order of a month or less. Here M is the mass of air within the vortex in a lower-stratospheric layer, say a density scale height or about 7 km deep, and $\dot{M}$ the average mass-flow rate at which the air within the vortex in the same layer is exported to middle latitudes, and replaced by downward flow within the vortex. Average rate for this purpose means averaged over the winter season.

Although the flowing processor hypothesis seems to be at variance with the results of fluid dynamical studies done so far, the possibly serious implications compel a careful re-examination of the basis for our present understanding of the fluid dynamics. One concern is that computer limitations preclude any direct check from a three-dimensional model having adequate spatial and temporal resolution. It is conceivable, perhaps, that small-scale, numerically unresolved motions might change the picture; so we need to bring to bear all available fluid-dynamical insight as well as modelling. The remainder of this paper presents a summary of such a re-examination; it is hoped to publish the results in more detail during the coming year. The conclusion will be that some vortex air is indeed exported, but at a rate that is both highly variable, and far less than one vortex mass per month.

3. The possibility of mean outflow

Any large export of vortex air in the lower stratosphere can be assumed to take place mainly along the isentropic surfaces of the strong stable stratification. This is because the most typical ways in which air can cross isentropic surfaces outside the vortex are quasi-diffusive, with small diffusivities K_{zz} of order $0.2 \text{ m}^2 \text{ s}^{-1}$ giving diffusion or dispersion height scales $(K_{zz}t)^{1/2}$ of only a kilometre or so for $t = 2$ months. The two main mechanisms (both giving small-step random walks in the cross-isentropic motion) are, first, local vertical mixing by clear-air turbulent layers

(Dewan 1981) and, second, quasi-horizontal, stratification-constrained (layerwise-two-dimensional) turbulence ('surf zone dynamics') giving a *diabatic* random walk with, by accident, as it happens, diffusivities of roughly the same order of magnitude, $0.2\,\mathrm{m^2\,s^{-1}}$. Such values are corroborated by lidar observations of volcanic aerosol layers (P. H. Haynes, personal communication). Superposed on these vertical random walks can be weak Lagrangian-mean descent rates, not well known but almost certainly of order perhaps 0.1–$0.2\,\mathrm{mm\,s^{-1}}$ or $\frac{1}{2}$–$1\,\mathrm{km}$ in 2 months. So any air rapidly exported from the vortex in the lower stratosphere would have to stay fairly close to one isentropic surface.

Transport along isentropic surfaces can be regarded as due either to mean outflow or to outward eddy transport, or to both. First consider the possibility of mean inflow or outflow along isentropic surfaces, as seen in vortex-following coordinates. Wave–mean interaction theory, together with observational knowledge of the relevant Rossby, gravity and inertia–gravity waves, points not to mean outflow but to weak mean inflow (Mo *et al.* 1995). This is part of the wave-driven gyroscopic pumping action already mentioned.

Additional weak inflow is needed to create the vortex on the seasonal timescale. Total poleward parcel displacements of a few degrees latitude are enough to create a vortex of the typical strength observed in the lower stratosphere, if drag on the vortex can be neglected; these displacements are robustly of the same order as would be given by a simplistic angular momentum budget for a frictionless, exactly circular vortex. Thus a hypothetical ring of air, conserving its angular momentum and initially at rest relative to the Earth, will be spun up to $30\,\mathrm{m\,s^{-1}}$ eastwards on arriving at latitude $60°$ if displaced polewards by only $2.1°$ latitude. These are typical lower-stratospheric vortex-edge values.

Essentially the same calculation run backwards implies that a mean outflow along isentropes strong enough to conform to the flowing processor hypothesis in the form quoted below (1), i.e. an outflow strong enough to export one vortex mass per month, would, in the absence of an eastward force, obliterate the vortex in 4 days or so. The area enclosed by a ring moving $2.1°$ equatorward *from* latitude $60°$ expands by a factor

$$\int_{57.9°}^{60°} \cos\phi\,\mathrm{d}\phi \Big/ \int_{60°}^{90°} \cos\phi\,\mathrm{d}\phi, \tag{2}$$

which is very close to $\frac{2}{15} = 4\,\mathrm{day}/1\,\mathrm{month}$. Similar estimates can be derived in several other, more sophisticated ways, for instance from wave–mean interaction theory (Mo *et al.* 1995) and from much more general considerations of the 'dilution of PV-substance' (Haynes & McIntyre 1990), which for this purpose apply to a disturbed as well as to an undisturbed vortex. But the orders of magnitude are little affected: 'strong eastward force of unknown origin' means very strong indeed: strong enough to spin up a new vortex every 4 days or so.

There is no known mechanism that could come anywhere near exerting an eastward (anti-drag) force as strong as this. The known forces, Rossby and gravity wave drag, are far weaker and in any case tend to be westward rather than eastward, being part of the gyroscopic pumping action.

In summary, weak mean inflow is required to create the vortex on the seasonal timescale. Additional mean inflow, of a similar order of magnitude, is generated by gyroscopic pumping action associated with Rossby and gravity wave drag. This inflow, induced by pumping, can also be looked on as maintaining the vortex against

Rossby and gravity wave drag. Conversely, if a strong mean outflow were to exist in the real wintertime polar lower stratosphere, it would imply an exceedingly strong *reversed* pumping action, driven by an unknown agency in the form of an exceedingly strong eastward force in the vortex edge.

4. Outward eddy transport by vortex erosion

If the foregoing is accepted, it leaves outward eddy transport as the only mechanism that could act against the presumed mean inflow. But outward eddy transport is rate-limited by the rate at which the vortex edge can be eroded, or material 'stripped' or 'peeled off', by disturbances to the edge and associated midlatitude stirring. The aforementioned model studies – which use either eulerian (fixed-grid) techniques or high-resolution lagrangian advection techniques, either contour advection or many-particle, on model-generated wind fields or meteorologically analysed wind fields (e.g. Pierce & Fairlie 1993; Manney *et al.* 1994*b*; Norton 1994; Rood *et al.*, personal communication; Fisher *et al.* 1993; Waugh & Plumb 1994; Waugh *et al.* 1993) – all give weak erosion rates, in the sense that the mass transported is, conservatively, no more than about a third of a vortex mass per month, regardless of the ambiguity in defining the vortex edge (due to its filamentary fine structure) and regardless of the very wide range of model resolutions and the consequent values of artificial model eddy diffusivities.

5. Outward eddy transport by inertia–gravity wave motions

The single effect neglected in these model studies is the possible parcel dispersion by the inertia–gravity wave field in the real lower stratosphere (Pierce *et al.* 1994). This should be roughly equivalent to a quasi horizontal Fickian diffusivity (McIntyre & Pinhey 1995). The reason is that the intermittent breaking of these waves by Kelvin–Helmholtz instability will give rise to a small-step quasi-horizontal random walking of the typical molecules of any chemical tracer along, as well as across (Dewan 1981), isentropic surfaces. The conclusion is unaltered even if the vortex edge has a fine-scale screwthread structure, as was suggested schematically in figure 1, making vertical mixing by breaking inertia–gravity waves (Dewan 1981) possibly significant as well as quasi-horizontal mixing along isentropes by the same waves. Estimates of the effective diffusivities in the vicinity of the vortex edge put it well within the range of artificial diffusivities used in the numerical model studies of vortex dynamics, and strongly reinforces the conclusions about limited vortex erosion rates already drawn from those studies. McIntyre & Pinhey (1995) look at this question systematically.

It can be added that, in order to sustain large transport rates down through, as well as out from, the vortex, of the order of one vortex mass per month, diabatic descent rates within the lower-stratospheric vortex would need to be several times greater than seems compatible with observed temperatures and with very extensive studies in atmospheric radiation, whose physics is fundamentally well understood, (e.g. Valero *et al.* 1993 & references). The predictions of such radiative studies receive independent support from direct, balloon-borne observations of descent of nitrous oxide isopleths within the Arctic polar vortex (Bauer *et al.* 1994). Nitrous oxide is chemically inert in the polar vortex and provides an excellent passive tracer of air motion. In the Antarctic diabatic descent can be expected to be, if anything, weaker than in the Arctic.

6. The stratospheric sub-vortex as a possible flowing processor

There is a transition altitude, usually at about 400 K, that appears to mark a transition between the stratospheric vortex and what will be called the stratospheric *sub-vortex* below. Like the vortex itself, the sub-vortex is also a region of low temperatures and dehydration, during late winter in the Southern Hemisphere and intermittently during winter in the Northern Hemisphere. But it is very different from the viewpoint of transport. It is also more massive because of the lower range of altitudes, below around 70 hPa. The distinction between the vortex and sub-vortex has been observationally clear for some time, from their different chemical signatures (Tuck 1989), and is consistent, moreover, with robust fluid-dynamical expectations based on the concept of 'PV inversion' (Hoskins *et al.* 1985).

The key point is that extratropical, synoptic-scale weather systems linked to strong, synoptic-scale PV anomalies on isentropic surfaces intersecting the tropopause, associated with distortions in the shape of the tropopause itself, have upward extensions into the stratosphere that tend to evanesce with altitude over height scales of a few kilometres. This upward evanescence is a robust property of the zonally asymmetric contributions to the wind, temperature and associated potential fields resulting from the inversion of synoptic-scale PV anomalies localized near the tropopause, mathematically similar to calculating the electrostatic potential due to localized charges. Such near-tropopause PV anomalies are very often part of the typical 'tropospheric' weather systems resulting from synoptic-scale cyclogenesis and anticyclogenesis in the extratropics; the PV anomalies near the tropopause tend to be strong, hence dynamically important, simply because isentropic gradients of PV tend to be strong near the tropopause. (Indeed it has long been found convenient, for reasons discussed in the review by Holton *et al.* (1995), to define the extratropical tropopause in terms of maximal isentropic gradients of PV; to this extent it is somewhat like the polar vortex edge.) The upward evanescence implies, then, that it is the lowermost isentropic surfaces in the lower stratosphere that are the most strongly stirred by the layerwise-two-dimensional motion induced by the synoptic-scale weather systems. Reinforcing this effect is the cyclostrophically dictated vertical shear of the vortex itself, making it weaker and more stirrable with decreasing altitude.

This is the overwhelmingly likely fluid-dynamical explanation for the existence of the transition altitude at about 400 K. According to this explanation (to which no alternative has been proposed, to my knowledge) the transition altitude is simply the altitude below which the layerwise-two-dimensional stirring is strong enough to overcome the Rossby quasi-elasticity of the vortex edge that would otherwise form. Any vortex edge that tries to form below the transition altitude, by diabatic or any other processes, is more or less stirred out of existence, depending on tropospheric weather activity. The same stirring is free to transport air parcels between the sub-vortex and the midlatitude regions on the same isentropic surfaces below the transition altitude. Such large-scale eddy transport can freely exchange polar and midlatitude air parcels without any requirement for a mean force field like that described in § 3; there is no single spinning air mass or coherent vortex to which angular-momentum and related arguments can be applied.

In this respect, the Arctic and Antarctic seem quite similar. Figure 2 clearly illustrates how such free exchange affects aerosols. The transition near 400 K is conspicuous.

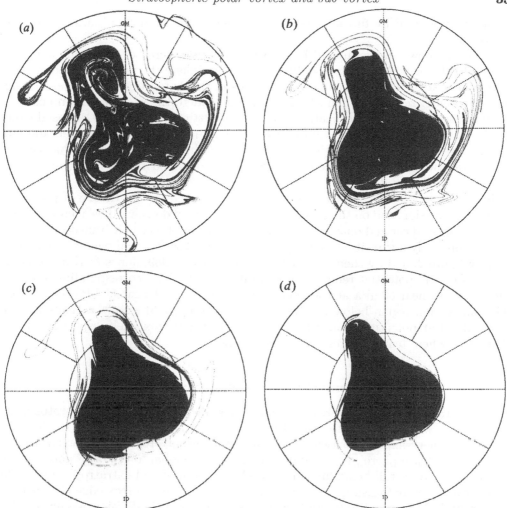

Figure 3. High-resolution tracer-advection pictures on different isentropic or stratification surfaces ((*a*) 350 K, (*b*) 375 K, (*c*) 400 K, (*d*) 425 K on 31 August 1993) illustrating the difference between Antarctic stratospheric vortex and sub-vortex behaviour, see text, as computed from meteorologically analysed winds over 40 days from 21 July to 31 August 1993 (Chen 1994). Maps are polar-stereographic out to 30° latitude, with the 60° latitude circle shown. A high-resolution adaptive Lagrangian 'contour advection' technique (Dritschel 1979; Norton 1994; Waugh & Plumb 1994) is used to trace the material contour that lengthens least hence best represents the vortex edge. The sub-vortex may well be more strongly ventilated than suggested by (*a*), in some winters at least, both because of the tendency of meteorological analyses to underestimate synoptic-scale, weather-related disturbances in data-sparse regions, and also because of the interannual variability of tropospheric weather-related disturbances beneath. In the Arctic, the sub-vortex may be even more strongly ventilated, consistent with its slightly higher vertical penetration (415 K as opposed to the Antarctic's 400 K (Proffitt *et al.* 1990)), and consistent with the aerosol distribution illustrated in figure 2.

An important corollary is that the sub-vortex could act as a dehydration and chlorine-activation site through which large masses of air could be efficiently transported, to and from middle latitudes at any time during winter. Such transport is indeed likely, since it would be brought about by the same large-scale eddy exchange processes that account for the existence of the transition altitude. This picture is strongly supported by further analyses of chemical tracer data (Jones & Kilbane-Dawe 1995).

Figure 3 shows an example of how vortex isolation increases with altitude, from Chen (1994), in an Antarctic case. In the real atmosphere the variation with altitude is probably stronger, since sparse observations in the Antarctic (and, in model studies, low model resolution) tend to result in an underestimation of the intensity of synoptic-scale tropospheric weather, hence of the stirring (Tuck 1994). Insight into the fundamental dynamics of Antarctic synoptic-scale tropospheric weather, and how it arises as baroclinic instabilities on a basic state driven, or strongly influenced by, the Antarctic katabatic surface drainage flow, is provided by an important recent paper by Juckes et al. (1994).

In summary, robust fluid-dynamical arguments based on the theory of potential-vorticity inversion, and on the cyclostrophically dictated downward weakening of vortex horizontal shear and quasi-elasticity, predict the existence of a transition altitude – more precisely a transition isentrope – which the observational evidence, exemplified by figure 2, and by chemical measurements (Tuck 1989; Jones & Kilbane-Dawe 1995), puts at potential temperatures around 400 K. The corresponding pressure altitudes are near 60 hPa at the centre of the vortex and near 80 hPa at its edge. The sub-vortex region below the transition isentrope could export activated air at a variable but potentially much larger rate than the vortex itself, especially in the Antarctic where temperatures are generally lower.

7. Postlude: the stratospheric overworld

By a strange accident, 400 K is also highly significant in the tropical stratosphere. Indeed, to a first approximation the 400 K isentropic surface divides the whole stratosphere rather simply into two very different transport regimes, an extratropical 'lowermost stratosphere', below about 400 K, and a global 'overworld' above about 400 K, the 'overworld' being further divided into tropical and extratropical parts by a subtropical eddy-transport barrier, weaker than the polar vortex edge but effective year round (Mote et al. 1995 and references therein). This division is another factor in the chemical differences seen above and below about 400 K, and shows more clearly the relevance of the extratropically pumped global-scale circulation to the problem of chemical 'stratosphere–troposphere exchange' (Holton et al. 1995).

I thank Dr Martin Wirth and Dr Ping Chen for kindly allowing me to reproduce figures 2 and 3, and Dr P. H. Haynes and Professor R. A. Plumb for important comments. This work received support from the Natural Environment Research Council through the UK Universities' Global Atmospheric Modelling Programme, from the Engineering and Physical Sciences Research Council through the award of a Senior Research Fellowship, and from the Isaac Newton Trust. The Centre for Atmospheric Science is a joint initiative of the Department of Chemistry and the Department of Applied Mathematics and Theoretical Physics.

Appendix A. Gyroscopic pumping versus global-scale heating

The fluid dynamics of the extratropical pumping action, and its interplay with infrared radiative relaxation, has been the subject of many careful analytical and numerical modelling studies and has been thoroughly reviewed by Holton et al. (1995). It depends on the abovementioned wavelike and turbulent eddy motions, which for this purpose can usefully be thought of in terms of breaking Rossby and gravity waves, with 'breaking' understood in a suitably generalized sense (McIntyre & Palmer 1985; McIntyre 1992, 1993a). These eddy effects give rise to a persistently one-signed, or ratchet-like, irreversible global-scale transport of angular momentum, which produces the global-scale pumping via an interaction with strong Coriolis forces that can be described as 'quasi-gyroscopic'. It is quasi-gyroscopic in the sense that pushing air, say, westwards tends to make it move polewards. For brevity's sake let us simply call it 'gyroscopic'.

This gyroscopic pumping involves, of course, the kind of non-local influence found in most fluid-dynamical problems. As with an indrawn breath, or the air sucked in by an ordinary domestic vacuum cleaner, the essential points are that constraints like mass conservation are crucial, and that associated influences may propagate via fast wave motions. For the indrawn breath, the relevant waves are ordinary acoustic waves. For the stratosphere, large-scale internal gravity and inertia–gravity waves are also relevant. The timescales of interest for global-scale chemical transport, say seasonal to decadal, are far longer than the wave propagation times in question. For practical purposes the concomitant non-local influence can be regarded as acting instantaneously. Further discussion of these points and of the modelling studies that demonstrate them is given in the review by Holton *et al.* (1995).

It is perhaps worth adding one further remark to that discussion. There is a widespread misconception that the rising branch of the global-scale stratospheric circulation is locally 'caused' by solar heating, as in a certain sense is true of the tropospheric Hadley circulation. But this is one of the ways in which the stratosphere differs from the troposphere. What is important for this purpose, in the stratosphere, is the far larger scale of the circulation and the far larger scale of the radiative heating and cooling, together with the relaxational character of the infrared contribution already mentioned. What this turns out to mean is that, under the conditions characteristic of the stratosphere, the main effect of solar heating is to raise temperatures rather than to cause any persistent global-scale circulation.

One reason for the misconception might be a tacit assumption that an equation of the form $A = B$, where B represents heating and A involves the circulation, implies a causal relation 'B causes A' notwithstanding that the equation could just as well be written $B = A$. (It is a similar misconception to conclude, from Newton's law 'fluid acceleration equals pressure-gradient force', that air moves toward the suction nozzle of a domestic vacuum cleaner 'because it is pushed by the pressure-gradient force'.) The similar misconception that solar heating 'causes' the global-scale stratospheric circulation tends, unfortunately, to be strengthened by the observed fact that the rising branch of the circulation has a slight tendency to follow the sun during the seasonal cycle. The rising branch leans slightly toward the summer side of the tropical upper stratosphere at solstice.

But the location of the rising branch, more or less within the tropical stratosphere, is enough in itself to demonstrate the incorrectness of the idea that the global-scale stratospheric circulation is caused by solar heating. This is because, throughout the greater part of the summer, the strongest diurnally averaged insolation and hence the strongest solar heating is nowhere near the tropics. Rather, it is at the summer pole. At solstice the insolation at the summer pole is, for instance, 39.5% higher than the average insolation over the tropical latitude band $\pm 20°$. A stratospheric circulation caused by, or dominated by, the pattern of solar heating would have to have its rising branch entirely over the summer pole at solstice, and not over the tropics at all.

It is easy to verify these points about insolation. For solar declination α relative to a spherical, rotating earth, the fractional length of day $\lambda(\phi)$ at latitude ϕ is

$$\lambda(\phi) = \pi^{-1} \arccos\left[\max\{-1, \min(1, -\tan\alpha\tan\phi)\}\right],$$

and the diurnally averaged vertical component of solar irradiance is the full solar irradiance multiplied by

$$S(\phi) = \lambda(\phi)\sin\alpha\sin\phi + \pi^{-1}\sin\{\pi\lambda(\phi)\}\cos\alpha\cos\phi.$$

It is easy to check first that the function $S(\phi)$ has an absolute maximum $S_{max} = \sin\alpha$ at the north pole, $\phi = 90°$, whenever α is within 2.8° of its maximum solstitial value $\alpha_{max} = 23.6°$, and second that, when $\alpha = \alpha_{max}$, $S_{max} = \sin\alpha_{max}$ is 1.395 times the area average of $S(\phi)$, i.e. the average weighted by a factor $\cos\phi$, over the tropical latitude band $\pm20°$.

Now visualize the declination α as 90° minus the angle between two vectors one of which gives the direction of the Earth's axis, i.e. points toward the pole star, and the other of which points from the Earth toward the Sun at the time of interest. Projecting this picture on the Earth's orbital plane, and approximating the Earth's orbit as circular, we get $\sin\alpha = \sin\alpha_{max}\cos\Delta t$, where Δt is the time after the northern summer solstice measured in units of $(1 \text{ year})/(2\pi)$. The time during which insolation is maximal at the summer pole is then, in units of 1 year,

$$\pi^{-1}\arccos\{\sin(\alpha)/\sin(\alpha_{max})\}$$

with $\alpha = 23.6° - 2.8°$, which is just over three-fifths of the summer season.

In summary, then, the relatively slight diversion of the rising branch of the stratospheric circulation toward the summer hemisphere shows just how weakly the stratospheric circulation pattern is influenced by solar heating, and how strongly by the gyroscopic pumping effects associated with the Earth's rapid rotation. It is precisely such gyroscopic pumping effects that single out the tropics as special rather than the summer pole. The seasonal cycle of the stratospheric circulation pattern is mainly due, therefore, to seasonality in the extratropical pumping.

Seen in this light, it is a strange coincidence – a perverse trick of Nature – that the circulation at higher, mesospheric altitudes should have a rising branch over the summer pole. That mesospheric circulation is in no way an exception to what has just been said. It is driven and controlled by gyroscopic pumping, due mainly to the violently breaking gravity waves in the lower thermosphere, above about 80 km. So strong is this pumping that the resulting refrigeration of the summer polar mesopause near 80–90 km makes it the coldest place on earth, as well as the sunniest place on earth, with temperature minima sometimes as low as 110 K. For further discussion the reader is referred to my 1992 review.

References

Bauer, R., Engel, A., Franken, H., Klein, E., Kulessa, G., Schiller, C., Schmidt, U., Borchers, R. & Lee, J. 1994 Monitoring the vertical structure of the Arctic polar vortex over northern Scandinavia during EASOE: regular N_2O profile observations. *Geophys. Res. Lett.* **21**, 1211–1214.

Charney, J. G. & Drazin, P. G. 1961 Propagation of planetary-scale disturbances from the lower into the upper atmosphere. *J. geophys. Res.* **66**, 83–109.

Chen, P. 1994 The permeability of the Antarctic vortex edge. *J. geophys. Res.* **99**, 20563–20571.

Dahlberg, S. P. & Bowman, K. P. 1994 Climatology of large-scale isentropic mixing in the Arctic winter stratosphere from analysed winds. *J. geophys. Res.* **99**, 20 585–20 599.

Dameris, M., Wirth, M., Renger, W. & Grewe, V. 1995 Definition of the polar vortex edge by LIDAR data of the stratospheric aerosol: a comparison with values of potential vorticity. *Beitr. Phys. Atmos.* (In the press.)

Dewan, E. M. 1981 Turbulent vertical transport due to thin intermittent mixing layers in the stratosphere and other stable fluids. *Science, Wash.* **211**, 1041–1042.

Fisher, M., O'Neill, A. & Sutton, R. 1993 Rapid descent of mesospheric air into the stratospheric polar vortex. *Geophys. Res. Lett.* **20**, 1267–1270.

Haynes, P. H. & McIntyre, M. E. 1990 On the conservation and impermeability theorems for potential vorticity. *J. Atmos. Sci.* **47**, 2021–2031.

Holton, J. R., Haynes, P. H., McIntyre, M. E., Douglass, A. R., Rood, R. B. & Pfister, L. 1995 Stratosphere–troposphere exchange. *Rev. Geophys. Space Phys.* (In the press.)

Hoskins, B. J., McIntyre, M. E. & Robertson, A. W. 1985 On the use and significance of isentropic potential-vorticity maps. *Q. Jl R. met. Soc.* **111**, 877–946. Corrigendum, etc. **113**, 402–404.

Jones, R. L. & Kilbane-Dawe, I. 1995 Observational evidence for the role of the stratospheric sub-vortex region in midlatitude ozone loss. *Geophys. Res. Lett.* (Submitted.)

Juckes, M. N. & McIntyre, M. E. 1987 A high resolution, one-layer model of breaking planetary waves in the stratosphere. *Nature, Lond.* **328**, 590–596.

Juckes, M. N., James, I. N. & Blackburn, M. 1994 The influence of Antarctica on the momentum budget of the southern extratropics. *Q. Jl R. met. Soc.* **120**, 1017–1044.

Junge, C. 1976 The role of the oceans as a sink for chlorofluoromethanes and similar compounds. *Z. Naturforsch.* **31a**, 482–487.

Lahoz, W. A. *et al.* 1995 Vortex dynamics and the evolution of water vapour in the stratosphere of the Southern Hemisphere. *Q. Jl R. met. Soc.* (In the press.)

Manney, G. L., Farrara, J. D. & Mechoso, C. R. 1994*a* Simulations of the February 1979 stratospheric sudden warming: model comparisons and three-dimensional evolution. *Mon. Wea. Rev.* **122**, 1115–1140.

Manney, G. L., Zurek, R. W., O'Neill, A. & Swinbank, R. 1994*b* On the motion of air through the stratospheric polar vortex. *J. atmos. Sci.* **51**, 2973–2994.

McIntyre, M. E. 1989 On the Antarctic ozone hole. *J. Atmos. Terrest. Phys.* **51**, 29–43.

McIntyre, M. E. 1992 Atmospheric dynamics: some fundamentals, with observational implications. In *Proc. Int. School Phys. 'Enrico Fermi' CXV Course. The Use of EOS for Studies of Atmospheric Physics* (ed. J. C. Gille & G. Visconti) ISBN 0444898964, pp. 313–386. Amsterdam: North-Holland.

McIntyre, M. E. 1993*a* On the role of wave propagation and wave breaking in atmosphere-ocean dynamics. In *Proc. XVIII Int. Congr. Theor. Appl. Mech., Haifa* (ed. S. Bodner, J. Singer, A. Solan, & Z. Hashin), 281–304. Amsterdam: Elsevier.

McIntyre, M. E. 1993*b* Isentropic distributions of potential vorticity and their relevance to tropical cyclone dynamics. In *Proc. ICSU/WMO Int. Symp. Tropical Cyclone Disasters, Beijing* (ed. J. Lighthill, Z. Zheng, G. Holland, K. Emanuel), 143–156. Beijing: Peking University Press.

McIntyre, M. E. 1994 The quasi-biennial oscillation (QBO): some points about the terrestrial QBO and the possibility of related phenomena in the solar interior. In *The Solar Engine and its Influence on the Terrestrial Atmosphere and Climate* (Vol. 25 of *NATO ASI Subseries I, Global Environmental Change*) (ed. E. Nesme-Ribes), 293–320. Heidelberg: Springer-Verlag.

McIntyre, M. E. & Palmer, T. N. 1985 A note on the general concept of wave breaking for Rossby and gravity waves. *Pure Appl. Geophys.* **123**, 964–975.

McIntyre, M. E. & Pinhey, N. J. G. 1995 On the possible effects of inertia–gravity waves on the permeability of the stratospheric polar-vortex edge. *J. geophys. Res.* (In preparation.)

Mo, R.-P., Bühler, O. & McIntyre, M. E. 1995 On the mean mass flux across the edge of a stratospheric polar vortex due to thermally dissipating, steady, nonbreaking Rossby waves. *Q. Jl R. met. Society.* (Submitted.)

Mote, P. W. *et al.* 1995 An atmospheric tape recorder: the imprint of tropical tropopause temperatures on stratospheric water vapor. *J. geophys. Res.* (Submitted.)

Norton, W. A. 1994 Breaking Rossby waves in a model stratosphere diagnosed by a vortex-following coordinate system and a technique for advecting material contours. *J. atmos. Sci.* **51**, 654–673.

Norton, W. A. & Chipperfield, M. P. 1995 Quantification of the transport of chemically activated air from the Northern Hemisphere polar vortex. *J. geophys. Res.* (In the press.)

Pierce, R. B. & Fairlie, T. D. 1993 Chaotic advection in the stratosphere: implications for the dispersal of chemically perturbed air from the polar vortex. *J. geophys. Res.* **98**, 18589–18595.

Pierce, R. B., Fairlie, T. D., Grose, W. L., Swinbank, R. & O'Neill, A. 1994 Mixing processes within the polar night jet. *J. atmos. Sci.* **51**, 2957–2972.

Tuck, A. F. 1989 Synoptic and chemical evolution of the Antarctic vortex in late winter and early spring, 1987. *J. geophys. Res.* **94**, 11687–11737. Corrigendum **94**, 16855.

Tuck, A. F. 1994 The use of ECMWF products in stratospheric measurement campaigns. In *Proc. Workshop on the Stratosphere and Numerical Weather Prediction*, pp. 73–105. Shinfield Park, Reading: European Centre for Medium-range Weather Forecasts.

Tuck, A. F., Russell, J. M. & Harries, J. E. 1993 Stratospheric dryness: antiphased dessiccation over Micronesia and Antarctica. *Geophys. Res. Lett.* **20**, 1227–1230.

Valero, F. P. J., Platnick, S., Kinne, S., Pilewskie, P. & Bucholtz, A. 1993 Airborne brightness temperatures of the polar winter troposphere as part of the AASE II and the effect of brightness temperature variations on the diabatic heating in the lower stratosphere. *Geophys. Res. Lett.* **22**, 2575–2578.

Waugh, D. W. & Plumb, R. A. 1994 Contour advection with surgery: a technique for investigating fine scale structure in tracer transport. *J. atmos. Sci.* **51**, 530–540.

Waugh, D. W. *et al.* 1993 Transport of material out of the stratospheric Arctic vortex by Rossby wave breaking. *J. geophys. Res.* **99**, 1071–1078.

Ozone loss in middle latitudes and the role of the Arctic polar vortex

By J. A. Pyle

Centre for Atmospheric Science, Department of Chemistry,
University of Cambridge, Lensfield Road, Cambridge CB2 1EW, UK

There have been a number of different suggestions as to the cause of the observed ozone decline over middle latitudes. Here, the particular impact of polar processes on the middle latitude lower stratosphere is discussed. Recent studies suggest that air, recently activated and then torn from the edge of the polar vortex, contributes to the observed ozone decrease. For example, observational and modelling studies both indicate that there is an important role for filaments of vortex air being stripped away from the vortex edge. However, there appears to be little support for the idea of the vortex as a massive 'flowing processor' through which large quantities of air, primed for ozone destruction, are transported.

1. Introduction

The cause of polar ozone depletion is now reasonably well understood: at low temperatures in the winter polar lower stratosphere, polar stratospheric clouds (PSCs) form. Reactions on the surface of the PSCs turn chlorine compounds from inactive forms (e.g. HCl, $ClONO_2$) into active forms (e.g. ClO) which, in the presence of sunlight, destroy ozone.

Much more controversial is the cause of the decline in ozone in middle latitudes of the Northern Hemisphere, amounting to about 4% per decade (annual average) since 1979 (WMO 1995). Most theories implicate the halogen compounds but the precise details of the processes involved remain in dispute. One possibility is that the ozone depletion occurs in polar latitudes and this air is then mixed into middle latitudes causing a general 'dilution' of ozone levels there. A second possibility is that air is primed for ozone depletion by the reactions on PSCs in polar regions, but is then transported southward, and possibly mixed, before the depletion occurs. If this process operates continuously (like a 'flowing processor', see McIntyre (this volume)) then large ozone loss might occur in mid-latitudes. A further possibility is that the chlorine is activated *in situ* in middle latitudes, possibly on sulphate aerosol, followed by local ozone depletion.

It is clearly important to establish quantitatively the degree of mixing between the polar vortex and lower latitudes. Large, rapid flow through the polar vortex has the potential to carry air to middle latitudes which could produce significant *in situ* ozone depletion. Furthermore, this process would itself be one of the major factors determining the chemical structure of the middle latitude lower stratosphere (and, incidentally, would imply that previous assessments using two-dimensional models of,

41

for example, the impact of aircraft on the lower stratosphere could be considerably in error).

There has been an energetic debate with a spectrum of views, ranging from the idea of the vortex as a continuous 'flowing processor' to the idea that the vortex is highly isolated and only mixes with middle latitudes (leading to 'dilution') at the time of the final warming. Unfortunately, the debate has been hampered by inadequate, and often differing, definitions of what is meant by the edge of the polar vortex, and so on. Nevertheless, there is a surprisingly large amount of objective agreement to be found in the literature. Many studies indicate that there can be considerable exchange between the edge of the vortex, where air can be chemically processed, and middle latitudes.

It is clearly of importance that quantitative studies of the extent of mixing are carried out using independent datasets, including some of the global fields available from the upper atmosphere research satellite (UARS). Special attention should also be paid to the region below 400 K which is generally more difficult to observe from space and is not probed so effectively by *in situ* aircraft measurements.

In the next section we summarize some data and modelling studies which have discussed the impact of polar processes on middle latitudes.

2. The impact of polar processes on middle latitudes

A number of studies using data, models, and a combination of both data and models have been carried out to investigate the extent to which air is mixed between the polar vortex and middle latitudes. Many of these have concentrated on the arctic vortex, studied extensively in the polar campaigns: the European Arctic stratospheric ozone experiment (EASOE); the second European stratospheric Arctic and middle latitude experiment (SESAME) and the airborne Arctic stratospheric expedition II (AASE II). Studies using UARS data have also appeared.

The results from some of these studies are presented below. The studies chosen all show mixing, to a greater or lesser extent, but none support the idea of a large and continuous flow through the polar vortex.

The erosion at the vortex edge has been demonstrated quite beautifully in a number of new studies using the technique of contour advection with surgery (Norton 1993; Waugh 1993; Plumb *et al.* 1994). Results from these studies show thin filaments being dragged around the vortex edge and being carried into middle latitudes. Interesting examples were reported during a disturbed period in January 1992. These results are consistent with the large scale potential vorticity (PV) maps of the Northern Hemisphere, but reveal structure on small scales which is only hinted at in a tantalising fashion in the meteorological analyses. That some of the structure is indeed real has been demonstrated by the very elegant four-dimensional variational assimilation work of Fisher *et al.* (1993).

Large scale model calculations are also consistent with these results. For example, Rood *et al.* (1992), Chipperfield *et al.* (1994) and Chipperfield (1994) have all described studies in transport models using analysed wind and temperature fields. Rood *et al.* (1992) conclude that intense cyclonic activity close to the vortex edge and large planetary scale events are the major mechanisms of extra-vortex transport. Nevertheless, in their study of a disturbed period in January and February 1989, only a small amount of processed air is found outside the polar vortex.

Chipperfield (1994) has studied the Arctic winters of 1991/92 and 1922/93. He considered the distribution of a model tracer, which records the number of hours of sunlight experienced by an air parcel which has also seen PSC temperatures within the previous 14 sunlight-days. The tracer contains the two conditions necessary for rapid ozone loss, PSC processing (to convert chlorine species to ClO) and exposure to sunlight (to enable the catalytic ozone destruction cycles to operate). Most of the modelled tracer is found well within the polar vortex but there is a considerable amount of tracer at about 160° E in a region of lower PV moving away from the vortex edge. A chemical model for the same period shows elevated ClO in this same region (Chipperfield *et al.* 1994). These results are in extremely good qualitative agreement with the contour advection results presented by Plumb *et al.* (1994). Thus, like the Rood *et al.* (1992) study, this model also shows mixing at the vortex edge. Despite being run for rather disturbed dynamical periods, neither model shows a large degree of continuous vortex to extra-vortex transport.

A number of other trajectory studies of the Northern Hemisphere support these conclusions. MacKenzie *et al.* (1994) looked at ensembles of isentropic trajectories, appropriately labelled if PSC conditions had been encountered within the previous 20 days. For both 1988/89 and 1991/92 they show that the midwinter vortex was filled with air which had been PSC-processed. Outside the vortex, they also found examples of processed air. However, in all these cases the active temperatures had been encountered at PV values characteristic of the vortex edge. They found no evidence of air having been ejected from the centre of the vortex.

Pyle *et al.* (1994) have carried out a case study of mid January 1992 during which adiabatic cooling in the stratosphere between Greenland and northwest Europe led to a large region favourable for PSC formation on and near the edge of the polar vortex. Using winds from a high resolution GCM integration initialized on 18 January, a large number of trajectories were started covering the area of low temperatures. Most of the trajectories remained close to the vortex edge, but a substantial number of air parcels were found away from the vortex at around 45° N close to the Caspian Sea, coincident with a region of high PV in the meteorological analyses for the same day. It seems clear that the feature is real, and at least in part arises from vortex erosion, probably exceptionally strong erosion in view of the exceptionally strong disturbance to the vortex.

In order to investigate the associated chemistry in more detail, a chemical package was integrated along a subset of the trajectories. For those trajectories staying close to the vortex edge, ozone depletions of around 1% over ten days were calculated. For the trajectories that moved to lower latitudes, a depletion of the order of 1% per day was calculated. It is clear that middle latitude depletion can arise from this process.

The studies discussed above have generally concentrated on relatively short periods or specific synoptic situations. In an attempt to draw some more general conclusions, Dahlberg & Bowman (1994) have carried out isentropic trajectory studies for nine northern hemispheres. Their conclusions, in broad agreement with the above, is that a barrier inhibiting mixing typically forms near the vortex boundary and is strongest in January and February. At 450 K the transport that does occur across the barrier is predominantly in the form of thin filaments ejected from the vortex. In December and March the mixing barrier is weaker due to non-conservative factors during the spin-up and breakdown (leading to dilution) of the vortex, respectively.

A number of these studies have also discussed occasions on which air is mixed from outside into the vortex. For example, Plumb *et al.* (1994) show evidence for transport into the Arctic vortex in January 1992, a case also mentioned by Pyle *et al.* (1994). In addition, Dahlberg & Bowman (1994) give a number of examples of poleward mixing, but indicate that these cases are generally rare. Presumably, as in January 1992, they are associated with a strongly disturbed polar vortex.

Finally, Pyle *et al.* (1995) have run some very high spatial resolution model studies (about $1° \times 1°$ in the horizontal) for both the SESAME and ASHOE (Antarctic and Southern Hemisphere ozone expedition) campaigns. They found excellent agreement between models and data on a number of occasions when vortex air (characterized, for example, by elevated active chlorine concentrations) was found outside the vortex. Their study demonstrates conclusively that air is removed from the vortex edge and can contribute to the ozone decline found in middle latitudes.

3. Summary

Our understanding of polar ozone has advanced dramatically since the discovery of the ozone hole in 1985. The advances in modelling capability and in the quality of atmospheric data since then offer a tremendous opportunity for understanding the new problems which have emerged, including the question of the connection between polar processes and middle latitudes. It is clear that the Arctic vortex can become every bit as perturbed chemically as its Southern Hemisphere counterpart. Chemical depletion of ozone undoubtedly occurs in the north. During the cold stratospheric winter of 1994/95, SESAME scientists reported ozone decreases of up to 50% within the polar vortex between 12 and 20 km (press release issued by DGXII of the European Commission on behalf of SESAME scientists, 30 March 1995). This appears to be the largest loss seen so far in the north. As stratospheric chlorine levels decrease early next century, in response to international regulations, the ozone loss should decline.

Attention is now focused on middle latitudes. Observational studies show that air, primed for ozone depletion, can be stripped from the vortex edge into middle latitudes. Models are also being used in high spatial resolution integrations to study the mixing process. The filaments of air, torn from the vortex edge, have now also been modelled. Sometimes the filaments are thin, tube-like structures; on other occasions they appear to have a greater vertical extent. In either case they seem to provide a mechanism by which air, activated in polar regions, can be mixed into middle latitudes and contribute to the ozone loss there.

References

Chipperfield, M. P. 1994 A 3-D model comparison of PSC processing during the Arctic winters of 1991/92 and 1992/93. *Ann. Geophys.* **12**, 342–354.

Chipperfield, M. P., Cariolle, D. & Simon, P. 1994 A 3D transport model study of chlorine activation during EASOE. *Geophys. Res. Lett.* **21**, 1467–1470

Dahlberg, S. P. & Bowman, K. P. 1994 Climatology of large scale isentropic mixing in the Arctic winter stratosphere, *J. geophys. Res.* **99**, 20585–20599.

Fisher, M. A., O'Neill, A. & Sutton, R. 1993 Rapid descent of mesospheric air into the stratospheric polar vortex. *Geophys. Res. Lett.* **20**, 1267–1270.

MacKenzie, A. R., Knudsen, B., Jones, R. L. & Lutman, E. R. 1994 The spatial and temporal extent of chlorine activation by polar stratospheric clouds in the Northern Hemisphere winters of 1988/89 and 1991/92. *Geophys. Res. Lett.*, **21**, 1423–1426.

Norton, W. A. 1994 Breaking Rossby waves in a model stratosphere diagnosed by a vortex-following coordinate system and a technique for advecting material contours. *J. atmos. Sci.* **51**, 654–673.

Plumb, R. A., Waugh, D. W., Atkinson, R. J., Schoeberl, M. R., Lait, L. R., Newman, P. A., Browell, E. V., Simmons, A. J. & Loewenstein, M. 1994 Intrusions into the lower stratospheric Arctic vortex during the winter of 1991/92. *J. geophys. Res.* **99**, 1089–1106.

Pyle, J. A., Carver, G. D. & Schmidt, U. 1994 Some case studies of chlorine activation during the EASOE campaign. *Geophys. Res. Lett.* **21**, 1431–1434.

Pyle, J. A., Chipperfield, M. P., Kilbane-Dawe, I., Lee, A. M., Stimpfle, R. M., Kohn, D., Renger, W. & Waters, J. W. 1995 Early modelling results from the SESAME and ASHOE campaigns. *Faraday Discuss.* **100**. (In the press.)

Rood, R., Douglass, A. & Weaver, C. 1992 Tracer exchange between tropics and middle latitudes. *Geophys Res. Lett.* **19**, 805–808.

Waugh, D. W. 1993 Contour surgery simulation of a forced polar vortex. *J. atmos. Sci.* **50**, 714–730.

WMO 1995 Scientific assessment of ozone depletion: 1994. *WMO Ozone Res. Monitoring Proj. Rep.* No. 37.

Solar irradiance, air pollution and temperature changes in the Arctic

BY G. STANHILL

Institute of Soils and Water, ARO, Bet Dagan 50-250, Israel

A highly significant decrease in the annual sums of global irradiance reaching the surface of the Arctic, averaging 0.36 W m^{-2} per year, was derived from an analysis of 389 complete years of measurement, beginning in 1950, at 22 pyranometer stations within the Arctic Circle. The smaller data base of radiation balance measurements available showed a much smaller and statistically non-significant change.

Reductions in global irradiance were most frequent in the early spring months and in the western sectors of the Arctic, coinciding with the seasonal and spatial distribution of the incursions of polluted air which give rise to the Arctic Haze.

Irradiance measured in Antarctica during the same period showed a similar and more widespread decline despite the lower concentrations of pollutants. A marked increase in the surface radiation balance was recorded. Possible reasons for these interpolar anomalies and their consequences for temperature change are discussed.

1. Introduction

Arrhenius (1896) first pointed out that any global warming resulting from increased atmospheric concentrations of CO_2 or other radiatively active gases will be enhanced in the polar regions. This has been confirmed by experiments with general circulation models in the equilibrium mode (Mitchell *et al.* 1990) although transient models indicate that the enhanced warming will be confined to the Arctic (Bretherton *et al.* 1990).

The importance of a substantial rise in the temperature of the Arctic is of wider significance than the local effects on the fragile ecosystems because of the many important, complex and interacting climatic feedback mechanisms involving the polar regions (Kellogg 1975).

Primary and secondary changes in the radiative exchange at the surface and in the atmosphere caused by the global increase in the concentration of radiatively active gases have been modified by incursions of polluted air reaching the Arctic from the industrial regions of the Northern Hemisphere.

In late winter and early spring, the influx and concentration of polluted, stagnant air give rise to the Arctic Haze phenomenon over large sectors of the region (Stonehouse 1986; Sturges 1991). Its effects on radiative exchange are complex, partly because they are strongly influenced by the shortwave reflectivity of the underlying surface. Calculations based on data gathered during research flights suggest that, in general, the Arctic Haze causes radiative cooling at the surface (Rosen & Hansen 1986; Valero & Ackerman 1986; Blanchet 1991).

Pollution-induced cooling, by offsetting the enhanced surface warming predicted for the Arctic by both equilibrium and transient general circulation models, could explain the absence of temperature increases in the Arctic demonstrated in several analyses of measurement series (Kukla & Robinson 1981; Kelley *et al.* 1982; Lamb 1982; Tsuchiya 1991; Jaworowski *et al.* 1992; Kahl *et al.* 1993; Walsh 1993).

Support for this explanation is provided in this study of changes in solar irradiance and net all-wave radiation balance at the surface which have been measured within the Arctic Circle during the last 40 years. The changes have been related to the seasonal and spatial patterns in air pollution and air temperature and contrasted with those derived from similar measurements in the Antarctic.

2. Changes in solar irradiance and radiation balance in the Arctic

(a) Data

Altogether 389 complete years of global irradiance $K\downarrow$ measurements were obtained from 22 sites. All-wave radiation balance Q_* was measured at 11 of these sites and 133 complete years of such data are available.

The coordinates of the measurement sites together with the periods covered are presented in table 1 with mean annual sums of $K\downarrow$, their interannual variations and cloudiness-indices – the fraction of extra-terrestrial irradiance. The references to data sources in most cases include details of the instruments used, their exposure and calibration.

The irradiance measurements analysed are all 24 h totals expressed either as mean monthly values or annual sums and corrected to the current World Radiation Reference scale (WMO 1981).

(b) Results

(i) Global irradiance

Annual sums of $K\downarrow$, averaged for each of the 43 years between 1950 and 1994 when data were available, are presented in figure 1. The calculated linear decrease of 0.36 ± 0.05 W m^{-2} per year was very highly significant, $p = 0.0001$; the addition of a quadratic term to the relationship did not increase its significance. The annual decrease of $K\downarrow$ calculated using the data from all sites, thus weighting each year for the number of measurements available, was 0.24 ± 0.06 W m^{-2} per year, less than that based on the average yearly values but equally significant, $p = 0.0001$.

Seasonal and annual trends in $K\downarrow$ at individual sites are given in table 2, expressed as the slope of linear regressions on the year of observation for those trends statistically significant at the 0.05 level or less; quadratic relationships yielded essentially similar results and therefore have not been presented.

Three-quarters of the Arctic sites showed significant linear trends in $K\downarrow$; one-quarter of both the monthly means and annual totals changed significantly. All but one of the significant annual changes were decreases; two-thirds of the significant monthly changes were negative in sign. In absolute terms, decreases in $K\downarrow$ were on the average twice as large as the fewer cases of increased irradiance.

Large seasonal and area differences in the distribution of the time trends in $K\downarrow$ were observed. One-third of all significant changes occurred in the two spring months, March and April, and one-half during the four months, February to May. Long-term trends in $K\downarrow$ were least frequent during the three midsummer months, June, July and August.

Table 1. *Global radiation measurements in the Arctic*

site	coordinates			period of measurement[a]	mean annual values $K\downarrow$ (GJ m^{-2})	F_E[b]	CV (%)	data sources[c]
Alert Canada	81°30′ N	62°20′ W	62 m	1964–93 (16)	2.897	0.51	7.4	AB
Krenkel former USSR	80°37′ N	58°03′ E	21 m	1964–89 (25)	2.691	0.47	5.1	A
Eureka Canada	80°00′ N	85°56′ W	10 m	1970–93 (9)	3.003	0.52	6.0	AB
Ny Alesund Norway	78°50′ N	11°30′ E	17 m	1974–92 (11)	2.373	0.41	5.9	C
Chelyuskin former USSR	77°43′ N	104°17′ E	12 m	1964–90 (22)	2.789	0.48	4.3	A
Mould Bay Canada	76°14′ N	119°20′ W	15 m	1965–87 (7)	2.954	0.50	4.8	AB
Kotelny Is. former USSR	76°00′ N	137°54′ E	11 m	1964–91 (22)	2.674	0.45	9.9	A
Resolute Canada	74°43′ N	94°59′ W	64 m	1964–93 (13)	3.150	0.53	5.3	AB
Bjornoya Norway	74°31′ N	19°01′ E	16 m	1970–93 (8)	2.066	0.35	7.3	D
Dickson Is. former USSR	73°30′ N	80°10′ E	42 m	1964–90 (22)	2.736	0.45	4.6	A
Sachs Harbour Canada	71°59′ N	125°17′ W	84 m	1970–86 (7)	3.233	0.52	6.0	AB
Barrow USA	71°08′ N	156°47′ W	19 m	1951–92 (14)	3.183	0.51	8.5	AEF
Wrangel Is. former USSR	70°58′ N	178°32′ W	2 m	1964–91 (24)	3.151	0.50	4.5	A
Chetyrekhstolbovoi former USSR	70°38′ N	162°24′ E	32 m	1964–91 (24)	3.238	0.52	4.8	A
Cambridge Bay Canada	69°06′ N	105°07′ W	23 m	1971–93 (7)	3.337	0.52	3.0	AB
Hallbeach Canada	68°47′ N	81°15′ W	7 m	1970–93 (8)	3.482	0.54	2.3	AB
Olenek former USSR	68°30′ N	112°26′ E	127 m	1964–91 (17)	3.251	0.50	5.8	A
Inuvik Canada	68°19′ N	133°32′ W	103 m	1950–93 (26)	3.380	0.52	4.5	ABF
Kiruna Sweden	67°50′ N	20°26′ E	408 m	1952–91 (22)	3.080	0.47	11.1	GF
Verkoyansk former USSR	67°33′ N	133°23′ E	137 m	1964–91 (22)	3.395	0.52	3.0	AI
Sodankyla Finland	67°22′ N	26°39′ E	138 m	1953–93 (36)	2.928	0.45	8.9	AF
Reykjavik Iceland	64°08′ N	21°54′ W	56 m	1958–91 (27)	2.897	0.42	8.5	AI

[a]Figure in brackets is the number of complete years. [b]F_E is the fraction of extraterrestrial. [c]See next page for data sources.

G. Stanhill

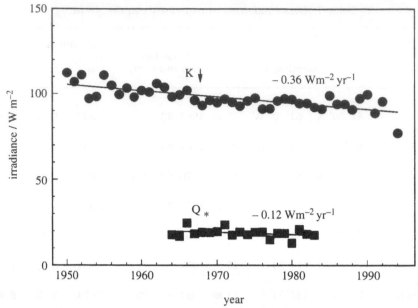

Figure 1. Changes in global irradiance and radiation balance within the Arctic Circle, 1950–1994. Average annual means for all complete years of measurements of the series listed in table 1.

The mean annual reduction in $K\downarrow$ within the Arctic Circle illustrated in figure 1 was 0.37% at individual sites. The average of annual reductions at individual sites was 0.15%. Decreases in monthly values of $K\downarrow$ were proportional to the irradiances and averaged 0.5% per year.

The area distribution of changes in $K\downarrow$, shown in figure 2, indicates reductions in the western sectors of the Arctic from Alaska to the Barents Sea; the small number of significant increases in irradiance were confined to central and eastern Siberia.

(ii) *Radiation balance*

Annual sums of Q_* averaged for each of the 21 years of data available between 1964 and 1984 are shown in figure 1. The calculated linear decrease of 0.12 W m^{-2} per year was not statistically significant.

The statistically significant trends found in monthly and annual values of Q_* at individual sites are presented in table 3 as the slope of linear regressions on year of measurement. As was the case with global irradiance, little statistical advantage was derived from substituting a quadratic in place of a linear relationship.

Ten out of the 11 Arctic measurement series had significant linear trends in Q_*; 29% of all changes in monthly means were significant, but only two of the 11 changes

Data sources for table 1. [A] Monthly Bulletins, Solar Radiation and Radiation Balance Data (The World Network), Voeikov Main Geophysical Observatory, St. Petersburg, Russia (since 1964). [B] Atmospheric Environment Service, Ontario, Canada. [C] Norsk Polarinstitutt Arbok (1974–9); Meddel Nr 118 (1992, 1981–87); V. Hisdal, personal communication (1994, 1988–92). [D] Norsk Meteorological Institute Arbok (1970–9); Norsk Meteorological Institute (1994, 1980–92). [E] Climate Monitoring and Diagnostics Laboratory NOAA, E. G. Dutton, personal communication (1994). [F] Marshanova and Chernigorskii (1978). [G] Sverges Meteorological and Hydrological Institute, personal communication, W. Josefson (1994). [H] Finnish Meteorological Service (1994). [I] Global Radiation in Iceland, M. A. Einarsson (1969); Iceland Meteorological Office (1994).

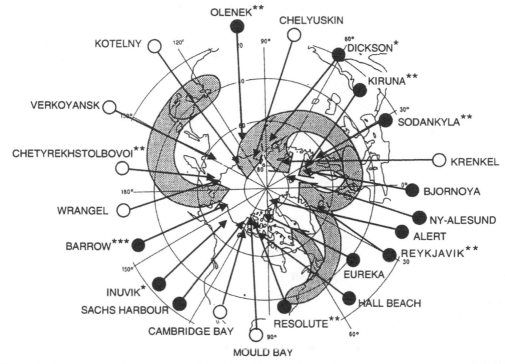

Figure 2. Mean annual changes in global irradiance within the Arctic Circle with location of pyranometer stations; closed circles indicate a decrease in irradiance, open circles an increase. The probability level of the fitted linear regression on year of measurement indicated by one star for $p = 0.05$, two for $p = 0.01$, and three for $p = 0.001$. Major sources and pathways for transport of pollutants between midlatitudes and the Arctic after Jaworowski (1989).

in annual sums were. Both of these trends in annual sums were negative, as were 60% of the significant trends in monthly values.

Neither the size nor the sign of the significant trends in Q_* were related to the size or sign of the radiation balance; the trends were of the same magnitude for increases and decreases.

The seasonal distribution of significant trends in Q_* was similar to that described for $K\downarrow$; 29% of the trends, nearly all of which were negative, occurred in March and April and 50% between February and May; only 3% of the significant changes occurred in September and October.

(c) Discussion

(i) Causes of reduction in global irradiance

The most common response to reports of large reductions in global irradiance is to question the accuracy of the measurements analysed. While instrumental errors cannot be excluded in individual cases, they are extremely unlikely to be the cause of the significant trends found in the large data base used in this study, which consists of measurements from a number of national meteorological services using a variety of pyranometer models of known accuracies, independently calibrated and operated. Many of the differences found in this data base exceed the random errors of measurement to be expected and the uncertainties of calibration procedures and of the standards on which these are based (WMO 1981).

G. Stanhill

Table 2. *Significant trends in global irradiance in the Arctic* $(\mathrm{W\ m^{-2}\ a^{-1}})$

site	I	II	III	IV	V	VI	VII	VIII	IX	X	XI	XII	year
Alert Canada		**0.01**	**−0.17**	**−0.94**					−0.38	*−0.02*			
Krenkel former USSR		*0.01*	0.24		*−0.79*			1.02					
Eureka Canada		0.01											
Ny Alesund Norway													
Chelyuskin former USSR		*0.02*					*0.71*						
Mould Bay Canada		*−0.04*	**−0.45**		*−1.18*				**−0.69**		0.01		
Kotelny Is. former USSR			*0.25*							0.14			
Resolute Canada		*−0.04*	**−0.47**	−0.72	**−1.43**	**−1.85**			*−0.29*	−0.11			−0.52
Bjornoya Norway			−0.67	−2.40									
Dickson Is. former USSR													
Sachs Harbour Canada	**−0.002**				2.88								
Barrow USA	−0.03	**−0.30**	**−0.54**	**−0.81**	−0.84	*−0.87*							**−0.23**
Wrangel Is. former USSR													
Chetyrkhstolbovoi former USSR													
Cambridge Bay Canada			−0.16	−0.37					−0.49	*−0.06*			
Hallbeach Canada													
Olenek former USSR					*1.45*					*−0.06*	*−0.02*		
Inuvik Canada	−0.03	**−0.13**	−0.27	**−0.60**	*−0.57*					*−0.14*	*0.01*	*−0.06*	
Kiruna Sweden	**0.06**	*−0.21*									**0.01**	**−0.06**	
Verkoyansk former USSR		*0.02*		*0.30*					*0.42*		**−0.03**		
Sodankyla Finland			−0.38				−1.07				*0.01*	*−0.11*	
Reykjavik Iceland							−1.69		0.17				−0.17

Bold values indicate significance at $P = 0.001$, plain at $P = 0.01$, *italic* at $P = 0.05$.

Changes in solar output, i.e. in the extra-terrestrial flux at the top of the earth's atmosphere, can also be excluded as the cause of the sometimes large but spatially variable reductions of $K\!\downarrow$ within the Arctic Circle. This is because of the spatial variation of the changes in $K\!\downarrow$, and also as the amplitude of the cyclic changes measured above the atmosphere are an order of magnitude less than the trends measured at the surface (Hartmann *et al.* 1993).

Increasing concentrations of radiatively active gases in the atmosphere can also be excluded as the cause of decreases in $K\!\downarrow$ as their prime effect on solar transmission will result from an increased water vapour concentration of the atmosphere following

Table 3. *Significant trends in radiation balance in the Arctic* (W m^{-2} a^{-1})

site	I	II	III	IV	V	VI	VII	VIII	IX	X	XI	XII	year
Alert Canada			**−0.75**				**−1.29**						
Krenkel former USSR			*−0.44*		*−0.88*		2.12				*−0.52*	*−0.54*	
Chelyuskin former USSR						**2.41**		*0.78*					
Mould Bay Canada			*−0.52*	*−0.89*			*−1.46*						-2.25
Kotelny Is. former USSR		**−1.20**	**−1.20**	**−1.76**	−1.56						−1.16	−0.97	**−0.82**
Resolute Canada	0.91	0.91	0.61	*−0.90*						**0.98**	**1.15**	**0.90**	
Dickson Is. former USSR		−0.41	−0.60	−0.75	*−1.09*		*−1.67*						
Wrangel Is. former USSR													
Chetyrekstolbovoi former USSR								*0.78*	**0.71**				
Olenek former USSR	**0.58**	0.41			*−2.23*						**0.65**	**0.57**	
Verkoyansk former USSR				*1.63*									

Bold values indicate significance at $P = 0.001$, plain at $P = 0.01$, *italic* at $P = 0.05$.

any heating which occurs. The simulated result of a doubling of CO_2 will be to reduce $K{\downarrow}$ by 5% (Wilson & Mitchell 1987).

Changes in the amount of cloud cover would have to be large to account for the reductions in $K{\downarrow}$ reported given the high transmissivity of Arctic cloud caused by their characteristically low water content and shallow depth (Gavrilova 1966). Using the mean value of transmissivity derived from three Arctic stations whose data are listed by Gavrilova, an annual increase of 0.4% in cloud cover is required to account for the average decrease in $K{\downarrow}$ of 0.36% a^{-1} shown in figure 1. An analysis of cloud cover observations made at seven Arctic stations during the last 50 years shows no evidence for any monotonic trends (Raatz 1981).

A substantial reduction in short wave irradiance reflected from the surface $K{\uparrow}$ would cause a decrease in $K{\downarrow}$ due to the close linkage between the two radiant fluxes under polar conditions (Hisdal 1982; Gardner 1987). Such a reduction could be caused by a lower surface reflectivity $K{\uparrow}\,K{\downarrow}^{-1}$ due to pollution or by a reduction in the area and duration of the Arctic sea ice cover. However, no evidence of a substantial or statistically significant interannual change – either in specific regions or in the total extent of Arctic sea ice – was found in an analysis of passive-microwave observations by satellites between 1978 and 1987 (Gloersen *et al.* 1992).

Aerosols brought into the Arctic atmosphere by incursions of polluted air from lower latitudes are almost certainly the major cause of the reductions in $K{\downarrow}$ reported in this paper. Their direct effect is to increase the scattering and absorption of short-wave radiation in the atmosphere and so reduce the flux transmitted to the surface. Indirectly, aerosol-cloud interactions modifying the microphysical properties of cloud and haze (Hobbs 1993) have the same effect. Another indirect and longer term effect of an increase in aerosol load, especially of carbon of anthropogenic origin, would result from their deposition on snow and ice surfaces, reducing their reflectivity and hastening their transformation into less reflective water surfaces. The results of five studies of the effects of Arctic aerosols on the solar irradiance absorbed at the surface have been tabulated by Blanchet (1991). Irradiance was reduced in all but one of the 19 model evaluations by amounts which ranged from $+0.8$ to -7.7 W m^{-2} and averaged -3.3 W m^{-2}. This corresponds to the measured mean reduction in Arctic $K\downarrow$ over a nine year period (figure 1).

Valero & Ackerman (1986) estimated the effect of Arctic haze on $K\downarrow$ at Barrow, Alaska, using a series of radiation profiles measured under cloud-free conditions in March 1983. They calculated a 20% reduction in net solar irradiance at the surface for both ice- and water-covered surfaces. This equals the reduction in $K\downarrow$ at Barrow measured between 1950 and 1983 (table 2).

(ii) *Causes of compensating changes in long-wave components of the radiation balance*

The radiation balance equation

$$Q_* = K\downarrow - K\uparrow + L\downarrow - L\uparrow,$$

where Q_*, $K\downarrow$ and $K\uparrow$ are as previously defined, and $L\downarrow$ and $L\uparrow$ are, respectively, the downwelling long wave atmospheric irradiance and the upwelling longwave terrestrial irradiance emitted from the surface, indicates that the 0.36 W m^{-2} mean annual reduction in $K\downarrow$ reported herein, must, in the absence of any significant change in the average value of Q_* (figure 1), be compensated either by increases in $L\downarrow$ or decreases in $L\uparrow$ and $K\uparrow$.

The increase in $L\downarrow$ due to the increased concentration of radiatively active gases in the atmosphere has been estimated to have averaged 0.06 W m^{-2} a^{-1} over the last decade (Shine *et al.* 1990).

The decrease in radiating surface temperature required to compensate fully for the averaged observed reduction in $K\downarrow$, is 0.07 °C per year, far greater than that observed in the long-term measurements series of Arctic surface temperatures analysed.

A decrease in $K\uparrow$ can be excluded as the cause of the reduction in $K\downarrow$ in the absence of evidence for a substantial decrease in the extent of ice cover, and hence short wave reflectivity, in the Arctic (Gloersen *et al.* 1992).

Thus the mechanism whereby the radiation balance at the Arctic surface has remained unchanged despite the decline in global irradiance, is not clear.

3. Changes in Antarctic irradiance

Increases in the surface air temperature of Antarctica measured during this century are among the highest recorded in the world (Braaten & Dreschkhoff 1992; Jacka & Budd 1992; Jones 1990; King 1994; Raper *et al.* 1984); this is in marked contrast both to the absence of warming in the Arctic and current predictions of transient global circulation models (Houghton *et al.* 1992).

This interpolar anomaly could be caused by the lower concentrations of pollutants in the southern polar regions due to the lesser industrial activity, population and land area of the Southern Hemisphere. The distribution of anthropogenic sulphate aerosols over the globe supports this explanation; their calculated atmospheric burden over the Southern Hemisphere is one-third of that over the Northern Hemisphere

(Charlson *et al.* 1991). A similar ratio between the two hemispheres was found in the reduction of $K\downarrow$ between 1958 and 1985: -0.44 W m^{-2} a^{-1} for the Northern and -0.14 W m^{-2} a^{-1} for the Southern Hemisphere (Stanhill & Moreshet 1992).

Pollutant levels measured in the two polar regions show a somewhat smaller but significant difference. The results of 16 studies tabulated by Heintzenberg (1989) show the concentration of particulates measured at the South Pole during the summer to be half of that measured in the Arctic. A similar ratio was found between the number of condensation nuclei and their rates of increase between 1974 and 1990, as measured continuously at the South Pole and at Barrow, Alaska (Kane 1994).

If pollution caused the reduction in $K\downarrow$ found in the Arctic, changes in Antarctic irradiance should be smaller in magnitude and spatially and seasonally more uniform than in the northern polar regions, because of the much longer pathways for anthropogenic aerosols reaching the southern polar regions. To test this hypothesis, the limited measurements of $K\downarrow$ and Q_* available for Antarctica were analysed.

(a) Data and results

Altogether 174 complete years of global irradiance measurements were obtained from ten sites; radiation balance was also measured at five of these sites, yielding 71 complete years of data. All but two of the measurement sites were at the coast between 70° W and 170° E. The two inland sites were at the South Pole at an elevation of 2800 m, and at Vostok (78°27′ S, 106°52′ E, 3488 m) near the Pole of Inaccessibility of the contiguous continent.

Annual sums of $K\downarrow$, averaged for each of the 34 years between 1957 and 1991 when complete years of data were available, decreased significantly, $p = 0.05$, with year of measurement; the calculated linear decrease was 0.33 W m^{-2} per year.

Significant reductions in $K\downarrow$ were found at three-quarters of the Antarctic sites; no increases were found. At the individual sites one-fifth of the monthly and annual reductions in $K\downarrow$ were statistically significant. No seasonal difference in the reductions of $K\downarrow$ were apparent when expressed relatively. The largest annual reductions in $K\downarrow$ were found in the eastern sector of the continent between 30° E and 170° E.

Annual sums of Q_* averaged for each of the 25 years between 1957 and 1987 for which complete years of data were available, increased significantly, $p = 0.01$, with year of measurement: the calculated linear increase was 1.03 W m^{-2} a^{-1}. By contrast, at individual sites increases in annual sums of Q_* were apparent only at Faraday and Novolazarevskaya, two of the four sites with long and complete series of measurements.

Monthly values of Q_* and $K\downarrow$ were highly correlated, $r > 0.9$, at all four sites, without the hysteresis in the seasonal relationship found at Arctic sites. At Novolazarevskaya the slopes and offsets of the linear seasonal relationship between Q_* and $K\downarrow$ increased from 1964 to 1987, reflecting the 25% increase in Q_* and 2.3% decrease in $K\downarrow$ which occurred.

4. Interpolar anomalies

The overall decrease in $K\downarrow$ measured in Antarctica, 0.33 W m^{-2} a^{-1}, is surprisingly large. It is greater than that of the Southern Hemisphere as a whole (Stanhill & Moreshet 1992) and similar to that found in the Arctic, despite the lower levels of pollution and higher levels of surface heating occurring in the southern polar regions.

The reduction cannot be attributed to a large or significant increase in cloud cover,

G. Stanhill

as no evidence for such a change was found in an analysis of observations from five Antarctic sites, including the continental radiation measurement station at Vostok, over the 1957–1985 period (Zav'ialova & Zhukova 1990). Nor is there any evidence for a large or significant decrease in the extent of Antarctic sea ice which, by reducing short-wave reflectance, could have caused a reduction in $K\downarrow$ (Gloersen *et al.* 1992).

The large overall increase in Q_* measured in Antarctica is even more surprising, although it is consistent with the anomalously rapid surface warming reported for the Antarctic. Moreover, this warming, averaging 0.028 °C per year at 15 coastal stations (Jacka & Budd 1992), would lead to a small decrease in Q_* through a greater emission of terrestrial irradiance from the surface.

The increase in Q_* is far greater than can be accounted for by the increase in long wave atmospheric radiation to the surface resulting caused by the greater concentration of radiatively active gases. Globally, this radiative forcing has increased by 0.97 W m^{-2} since 1960 and by 0.06 W m^{-2} a^{-1} during the last decade (Shine *et al.* 1990). These calculations do not, however, include the recently demonstrated effect of ozone depletion in reducing solar irradiance (Isaksen 1994).

Thus, in the Antarctic as in the Arctic, the cause of the changes in the components of the surface radiation balance are not clear: in part this uncertainty is attributable to the limited number and duration, as well as accuracy, of the measurement series available.

5. Conclusions

The magnitude of the reduction in global irradiance measured within the Arctic during the last four decades and its seasonal and spatial variation support the hypothesis that this decline was caused by incursions of polluted air. This reduced irradiance may, by compensating for radiative forcing due to the increased concentration of radiatively active trace gases, have prevented warming of the Arctic surface.

The absence of a corresponding decrease in the measured values of the surface radiation balance could not be explained by compensating changes in other components of the radiation balance.

Reductions in global irradiance measured in Antarctica were of similar magnitude to those found in the Arctic despite the lower level of pollutants in the southern polar region. This reduction in $K\downarrow$ was accompanied by a much larger increase in the surface radiation balance. Although this is compatible with the rapid surface warming reported in Antarctica, no explanation of these interpolar anomalies was found.

A better understanding of the changes in the polar radiation balances, which are of global as well as regional importance for climate change, will require a sustained program of measurements in which all of the components of the surface radiation and energy balances are monitored to the maximum accuracy currently attainable.

It is a pleasure to acknowledge the assistance of Shirley Sawtell of the Scott Polar Research Institute, UK, for help in locating the data used in this study. Thanks are also due to the meteorological services of Australia, Canada, Finland, Iceland, Norway, Sweden and the USA for supplying unpublished measurements; and to S. Cohen and S. Moreshet of the Agricultural Research Organization, Israel, for their help in the computations and preparation of the figures.

References

Arrhenius, S. A. 1896 On the influence of carbonic acid in the air upon the temperature of the ground. *Phil. Mag. S.* **41**, 237–276.

Blanchet, J. P. 1991 Potential climate change from Arctic aerosol concentration. In *Pollution of the Arctic atmosphere* (ed. W. T. Sturges), pp. 289–322. London: Elsevier.

Braaten, D. A. & Dresckhoff, G. A. M. 1992 Maximum and minimum temperature trends at McMurdo Sound station. *Antarctic J.* **27**, 282–283.

Bretherton, F. P., Bryan, K. & Woods, J. P. 1990 Time-dependent greenhouse-gas-induced climate change. In *Climate change: the IPCC scientific assessment* (ed. J. T. Houghton, G. J. Jenkins & J. J. Ephraums), pp. 173–192. London: Cambridge University Press.

Charlson, R. J., Langner, J., Rodhe, H., Leavy, C. B. & Waren, S. G. 1991 Perturbations of the Northern Hemisphere radiative balance by backscattering from anthropogenic aerosols. *Tellus* **43AB**, 152–163.

Einarsson, M. A. 1969 *Global radiation in Iceland*. Reykjavik: Iceland Meteorological Service.

Gardner, B. G. 1987 Solar radiation transmitted to the ground through cloud in relation to surface albedo. *J. Geophys. Res.* **92 D4**, 4010–4018.

Gavrilova, M. K. 1966 *Radiation climate of the Arctic*. Jerusalem: Israel Program for Scientific Translations IV. (Translated from the Russian: *Radiatsionnyi klimat Arktiki* 1963. Leningrad: Gidrometeorologischkoe Izdatel'stvo.)

Gloersen, P., Campbell, W. J., Cavelieri, P. J., Comiso, J. L., Parkinson, C. L. & Zwally, H. J. 1992 Arctic and Antarctic sea ice, 1978–1987. *NASA XXIX* **SP-511**.

Hartmann, D. L., Barkstrom, B. R., Crommelynck, D., Fonkal, P., Hansen, J. E., Lean, J., Lee, R. B. III, Shoeberl, M. R. & Wilson, R. C. 1993 Total solar irradiance monitoring. *Earth Observer* **5**(6), 23–27.

Heintzenberg, J. 1989 Arctic haze: air pollution in polar regions. *Rapportserie* Nr. 53. Oslo: Norsk Polarinstitutt.

Hisdal, V. 1982 Sun-sky and water-sky luminance at an Arctic station. *Polar Res.* **2**, 3–15.

Hobbs, P. V. 1993 Aerosol–cloud interactions. In *Aerosol–cloud–climate interactions* (ed. P. V. Hobbs), pp. 33–73. San Diego: Academic Press.

Houghton, J. T., Callender, B. A. & Varney, S. K. (eds) 1992 *Climate Change: supplementary report to the IPCC scientific assessment*. London: Cambridge University Press.

Isaksen, I. S. A. 1994 Dual effects of ozone reduction. *Nature, Lond.* **372**, 322–323.

Jacka, T. H. & Budd, W. F. 1992 Detection of temperature and sea ice extent changes in the Antarctic and Southern Ocean. In *Int. Conf. Role Polar Regions in Global Change: Proc. Conf. June 11–15, 1990 Univ. Alaska, Fairbanks* (ed. G. Weller, C. L. Wilson & B. A. B. Severin), vol. 1, pp. 63–70. Fairbanks: University of Alaska.

Jaworowski, Z. 1989 Pollution of the Norwegian Arctic: a review. *Rapportserie* Nr. 55. Oslo: Norsk Polarinstitutt.

Jaworowkski, Z., Segalstad, T. V. & Hisdal, V. 1992 *Atmospheric CO_2 and global warming: a critical review* (2nd rev. edn). *Meddelelser* Nr. 119. Oslo.

Jones, P. D. 1990 Antarctic temperatures over the present century – a study of the early expedition record. *J. Climatol.* **3**, 1193–1203.

Kahl, J. D., Charlevoix, D. J., Zaitseva, N. A., Schnell, R. C. & Serreze, M. C. 1993 Absence of evidence for greenhouse warming over the Arctic ocean in the past 40 years. *Nature, Lond.* **361**, 335–337.

Kane, R. P. 1994 Interannual variability of some trace elements and surface aerosol. *Int. J. Climatol.* **14**, 691–704.

Kelley, P. M., Jones, P. D., Sear, C. B., Cherry, B. S. G. & Tavokol, R. K. 1982 Variations in surface air temperatures. Part 2. Arctic regions 1881–1980. *Mon. Weath. Rev.* **110**, 71–83.

Kellogg, W. W. 1975 Climate change feedback mechanisms involving the polar regions. In *Climate of the Arctic* (ed. G. Weller & S. A. Bowling), pp. 111–116. Fairbanks: Geophysical Institute University of Alaska.

King, J. C. 1994 Recent climate variability in the vicinity of the Antarctic peninsula. *Int. J. Climat.* **14**, 357–369.

Kukla, G. J. & Robinson, D. A. 1981 Temperature changes in the last 100 years. In *Climate variations and variability: facts and theories* (ed. A. Berger), pp. 287–301. Dordrecht: D. Reidel.

Lamb, H. H. 1982 *Climate history and the modern world*, pp. 259–261. London: Methuen.

Marshunova, M. S. & Chermingovskii, N. T. 1978 Radiation regime of the foreign Arctic. VI. New Delhi: Indian National Scientific Documentation Centre. (Translated from the Russian: *Radiatsionnyi rezhum zarabezhnoi Arktiki* 1961. Leningrad: Gidrometeorologischkoe Izdatel'stvo.)

Mitchell, J. F. B., Manabe, S., Tokioka, T. & Maleshko, V. 1990 Equilibrium climate change. In *Climate change: the IPCC scientific assessment* (ed. J. T. Houghton, C. J. Jenkins, & J. J. Ephraums), pp. 131–178. London: Cambridge University Press.

Raatz, W. E. 1981 Trends in cloudiness in the Arctic since 1920. *Atmos. Environ.* **15**, 1503–1506.

Raper, S. C. B., Wigley, T. M. L., Mayes, P. R., Jones, P. D. & Salinger, M. J. 1984 Variations in surface air temperatures. 3. The Antarctic 1957–1982. *Mon. Weath. Rev.* **112**, 1341–1353.

Rosen, H. & Hansen, A. D. A. 1986 Light-absorbing combustion-generated particulates over reflecting polar ice. In *Arctic air pollution* (ed. B. Stonehouse), pp. 101–120. London: Cambridge University Press.

Shine, K. P., Derwent, R. G., Wuebbles, D. J. & Morcrette, J-J. 1990 Radiative forcing of climate. In *Climate change: the IPCC scientific assessment* (ed. J. T. Houghton, G. J. Jenkins & J. J. Ephraums), pp. 41–68. Cambridge: Cambridge University Press.

Stanhill, G. & Moreshet, S. 1992 Global radiation climate changes: the World Radiation Network. *Clim. Change* **21**, 57–75.

Stonehouse, B. (ed.) 1986 *Arctic air pollution*, vol. XVI. London: Cambridge University Press.

Sturges, W. T. (ed.) 1991 *Pollution of the Arctic atmosphere*. London: Elsevier.

Tsuchiya, I. 1991 Recent secular trends of surface air temperatures in the high latitudes of the Northern Hemisphere. *Proc. NIPR Symp. Polar Meteorol. Glaciol.* **4**, 43–51.

Valero, F. P. J. & Ackerman, T. P. 1986 Arctic haze and the radiation balance. In *Arctic air pollution* (ed. B. Stonehouse), pp. 121–133. London: Cambridge University Press.

Walsh, J. E. 1993 The elusive Arctic warming. *Nature, Lond.* **361**, 300–301.

Wilson, C. A. & Mitchell, J. F. B. 1987 A doubled CO_2 climate sensitivity experiment with a global climate model including a simple ocean. *J. Geophys. Res.* **92**, 13 315–13 343.

WMO 1981 Measurement of radiation. In *Guide to meteorological instrument and observing practice*, ch. 9. Geneva: World Meteorological Organization.

Zav'ialova, I. N. & Zhukova, O. L. 1990 Multiyear cloudiness variability in Antartica (in Russian). In *Meteorological research in the Antarctica. Collected Papers of the 3rd All-Union Symposium. Part I* (ed. Ye. S. Korotkevich), pp. 51–54. St. Petersburg: Gidrometeorologischkoe Izdatel'stvo.

Discussion

T. V. CALLAGHAN (*Centre for Arctic Biology, School of Biological Sciences, University of Manchester, UK*). Professor Stanhill has stated that there is no evidence of climatic warming so far in the Arctic. This observation conflicts with results from the collation of temperature data from official meteoriligical sations throughout the Arctic (Jones & Briffa 1992). Their definitive study has shown that some areas of the Arctic have experienced increases in annual mean temperature of up to 4.5 °C over the 30 year period 1960–1990. Jones & Briffa (1992) also show that other areas of the Arctic, such as Fennoscandia, have not experienced recent changes in temperature while the Baffin Bay–West Greenland region has sent recent cooling.

G. STANHILL. Thank you for drawing my attention to this valuable reference.

Additional references

Jones, P. D. & Briffa, K. R. 1992 Global surface air temperature variations during the twentieth century. Part 1. Spatial, temporal and seasonal detail. *The Holocene* **2**, 165–179.

Arctic terrestrial ecosystems and environmental change

By Terry V. Callaghan[1]† and Sven Jonasson[2]

[1]Centre for Arctic Biology, School of Biological Sciences,
University of Manchester, Oxford Road, Manchester M13 9PL, UK
[2]Department of Plant Ecology, University of Copenhagen,
Øster Farimagsgade 2D, DK-1353 Copenhagen K, Denmark

The impacts of environmental change on Arctic terrestrial ecosystems are complex and difficult to predict because of the many interactions which exist within ecosystems and between several concurrently changing environmental variables. However, some general predictions can be made.

(i) In the sub-Arctic, subtle shifts in plant community composition with occasional losses of plant species are more likely than immigration of exotic species. In the high Arctic, colonization of bare ground can proceed and there are likely to be shifts in ecotypes. Major shifts in vegetation zones, such as the advance of the boreal forest, are likely to be slow and species specific responses will result in different assemblages of species in plant communities in the longer term. All changes in community structure, apart from species removal by direct extreme weather conditions (e.g. drought) will be slow because of the slow growth, low levels of fecundity and slow migration rates of plant species over large latitudinal ranges.

(ii) Mobile mammals and birds can probably adjust to changes in the distribution of their food plants or prey in the Arctic, but vertebrate and invertebrate herbivores may face problems with changes in the quality of their food plants. Non-migratory animals could be severely affected by altered winter snow conditions which affect availability of food and shelter.

(iii) Increases in primary production are uncertain and depend mainly upon the responses of soil microbial decomposer activity to changes in soil temperature, moisture and plant litter quality. Assumptions that climate warming will lead to warmer soils and increased nutrient availability to sustain higher productivity are uncertain as greater biomass may lead to reduced soil temperatures through insulation effects and increased nutrients released may be immobilized by soil microorganisms.

(iv) Changes in environmental conditions are themselves often uncertain. There is particular doubt about changes in precipitation, growing season length, cloudiness and UV-B radiation levels while such environmental changes are likely to vary in magnitude and direction between different regions of the Arctic.

(v) The large populations and circumpolar distributions typical of Arctic biota lead to a strong buffering of changes in biodiversity. Perhaps the greatest threats to Arctic biota will be imposed by the degradation of permafrost which may lead to either waterlogging or drought depending upon precipitation regimes.

---

† Present address: Sheffield Centre for Arctic Ecology, Department of Animal and Plant Sciences, The University of Sheffield, Tapton Experimental Gardens, 26 Taptonville Road, Sheffield S10 5BR, UK.

1. Introduction

The interactions between Arctic terrestrial ecosystems and environmental change are receiving increasing attention (e.g. Chapin *et al.* 1992; Callaghan *et al.* 1992, 1995) for several reasons. Arctic terrestrial ecosystems cover more than 7 million km² and contain rich biological resources adapted to the particular environment there, together with mineral and fossil fuel resources. While biotic and abiotic resources of the Arctic are particularly important to the life support of native and immigrant peoples living there, biological processes contribute to the global carbon budget through the sequestration of large amounts of atmospheric carbon in tundra organic soils. The Arctic also contributes to the global energy balance through its relatively high albedo compared with that of the boreal forest (Dickinson & Hanson 1984).

Any change in Arctic terrestrial ecosystems is therefore, likely to have local, regional and possibly global impacts. Climate change is predicted by General Circulation Models (GCMs) to be greatest at high latitudes (Mitchell *et al.* 1990) while even small changes in temperature are likely to stimulate disproportionately large biotic responses in those organisms already close to their lower temperature limits for survival. Further, slow growth or development and low fecundity of terrestrial primary producers together with low species diversity of Arctic biota result in the fragility of Arctic ecosystems experiencing disturbance or environmental change. The degree of environmental change expected in the Arctic together with the possibility of disproportionately great impacts on the biota present a system in which early indications of environmental change can be detected, while the considerable longevity and slow response of many Arctic organisms offer a system in which the signal to noise ratio is high.

Environmental change is not new in the Arctic: plants and communities thriving in the cold period immediately before the Little Ice Age some 400–500 years ago on Ellesmere Island, now exist in a much warmer climate (Havström *et al.* 1995). Climate has even changed over the past 30 years with warming of 4.5 °C in some continental areas and cooling in others (Jones & Briffa 1992). However, it is a combination of natural environmental changes combined with numerous types of anthropogenic stresses and disturbances (e.g. greenhouse induced climate change, transboundary pollution leading to eutrophication, mining impacts, etc.) which now present particular threats to terrestrial Arctic ecosystems and a challenge for researchers aiming to predict and limit their ecological impacts.

2. Environmental characteristics of Arctic terrestrial ecosystems

(a) Stress

The Arctic is characterized by a long, dark, cold winter when water necessary for life processes is in solid form. Consequently, Arctic organisms have resting phases, migrate or are physiologically and morphologically adapted to withstand the winter. Summers with mean daily temperatures exceeding zero may be less than 6 weeks in the high Arctic but high temperatures may be experienced briefly and continuous sunlight allows some plant species to attain positive energy balances throughout the 24 hour cycle.

Precipitation is variable and soil moisture often varies from the extremes of flooding and anaerobic conditions during spring thaw to drought in arid polar regions during late summer (compare figure 3*a* with 3*c*). Topography, even at the microscale (cm) and permafrost dynamics are important determinants of soil moisture for plants, soil microbes and invertebrates and exposure is critical for plants, some birds and mammals. Soil moisture and soil and air temperature gradients can be extreme; plants can be 23.5 °C warmer than the air a short distance above (Mølgaard 1982). Such temperature differentials may lead to problems such as late winter desiccation in plants when shoots are warm and snow free but roots are frozen in the soil.

(b) Disturbance and mechanical impacts

Many tundra habitats are characterized by extreme stress and disturbance, the latter usually associated with permafrost dynamics, freeze–thaw cycles in soil and rock, and water and wind erosion. Degradation of permafrost may disturb whole landscapes, e.g. when the active layer becomes detached or during 'baidgerahk' formation (figure 3b). Freeze–thaw cycles produce patterning in the landscape. They disrupt plant communities and soil fauna when ice grows and hummocks, polygons, pingos and palsas form. However, such disturbance can be associated with a drying of the soil, an increase in soil temperature and an increase in biotic diversity as plant seedlings become established and rodents, foxes, wolves, snowy owls etc. inhabit the warm dry soil.

Water erosion, particularly during permafrost degradation, removes soil organic matter containing plant nutrients as well as underlying mineral soils carrying them onto the vast continental shelves of the Arctic. Disruption to plant and soil communities can be severe. Wind provides a mechanical impact on exposed organisms which have evolved to either escape or to reduce resistance in aerodynamically smooth forms such as cushion plants. Wind also erodes both substrates and plants, particularly through ice crystal abrasion during winter.

(c) Snow cover

The persistent snow cover limits the length of the active period for plants as the ground becomes snow-free first in late June or in July when the solar angle is already decreasing. However, the snow offers comparatively benign environments for plant meristems, hibernating invertebrates and winter-active small rodent herbivores in the sub-nivean space beneath the snowpack, and creates protection from larger herbivores and predators. The small rodents, common over most of the Arctic, forage in winter where snow accumulates, while large mammals, e.g. musk-oxen, caribou and dall-sheep, exploit habitats with thin snow cover.

(d) Resource status

Space shows an increasing limitation from the barren polar high Arctic deserts (figure 3a) to the completely vegetated shrub tundras of the low Arctic (figure 3e). Although plant nutrients may be locally abundant in Arctic soils, they are generally unavailable to plants (Shaver & Chapin 1980) and this resource limitation, generally more than the stresses and disturbance described above, results in low rates of primary production. Nutrient limitations of primary production usually result from low decomposer activities in cold, often anaerobic soils together with the sequestration of nutrients in microbial biomass. Low levels of primary production limit the productivity of other trophic levels dependent on the plants.

Atmospheric carbon dioxide, an important resource for plant photosynthesis, is increasing in concentration, particularly at high northern latitudes (Goreau 1990). However, it is unclear to what extent this resource alone limits primary production as some plant responses to elevated levels of carbon dioxide last only for weeks and whole ecosystem response is no longer observed after three years (Tissue & Oechel 1987).

Resource quality can also be an important limitation. Plants with high carbon to nitrogen ratios, for example those grown in high concentrations of atmospheric CO_2, and particularly those with secondary metabolites (e.g. those grown in high UV-B radiation), limit both rates of decomposition and rates of consumption by invertebrate and vertebrate herbivores (Bryant & Reichardt 1992).

3. Biotic characteristics

(a) Growth and life cycle strategies of plants

Plants generally are slow growing, long-lived and have low fecundity. Primary production rates and phytomass are low with a disproportionate amount of phytomass (up to 98%) in belowground structures. In closed vegetation of the sub-Arctic, clonal plants predominate and recruitment from seedlings is often restricted in both time and space. In the open vegetation of the high Arctic, recruitment by seedlings and viviparous propagules is more common but is often sporadic. Patchiness of habitat and resources is a dominant feature of Arctic landscapes (e.g. polygons (figure 3c) and hummocks (figure 3d)) and plant adaptations to these environments are common, for example a high degree of physiological integration between generations within a clone (Jonsdottir & Callaghan 1990). Physiological strategies have been claimed to be limited, with most plants adopting avoidance strategies or morphological and developmental adaptations. However, freeze tolerance is an obvious prerequisite for existence in the Arctic and adaptation to immediate recovery from anaerobic conditions has recently been shown to be important in relation to encapsulation in ice (Crawford *et al.* 1994). Also, photosynthetic optima are lower in Arctic plants than in their temperate counterparts.

(b) Characteristics of Arctic animals

Arctic lakes and ponds often have a high productivity of insects with aquatic larval stages and other invertebrates (e.g. crustaceans) that form the food source of higher aquatic or terrestrial animals. The numerous small ponds offer particularly good feeding places for several aquatic birds and for waders because some trophic levels containing predators of invertebrates, such as fish, may be absent, particularly in those shallow ponds in which water freezes to the bottom in winter. Thus, the Arctic contains abundant food and attracts migratory birds during the short summer – over 6 million North American sea birds nest in the Arctic regions of that continent while far greater numbers (mainly waders) migrate to the Siberian Arctic each summer. Most birds of the Arctic and some mammals are migratory species. A few species stay in the Arctic all year round and remain active above or below the snow during the entire winter period whereas others, as for instance most invertebrates and some larger mammalian species, enter diapause or hibernate, respectively.

(c) Biodiversity

Low fecundity of plants, long life spans and large latitudinal separation from biologically diverse latitudes together with the environmental filter of freezing have resulted in low species diversity among Arctic plants (figure 1). However, where sexual reproduction is more frequent in the high Arctic, subspecies diversity can be high

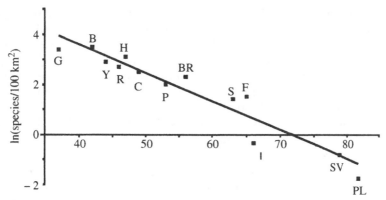

Figure 1. Biodiversity trends in vascular plants of Europe and the European Arctic. $r = 0.957$, $p < 0.001$; (G) Greece; (B) Bulgaria; (Y) Yugoslavia; (R) Romania; (H) Hungary; (C) Czechoslovakia; (P) Poland; (BR) Baltic Republics; (S) Sweden; (F) Finland; (I) Iceland; (SV) Svalbard; (PL) Peary Land, northeast Greenland. (Based on Hendry 1993 and Billings 1992.)

(Crawford *et al.* 1993). Fecundity of many birds and mammals is high although often cyclical as in rodents, which in turn also affects the reproduction of their predators. The large Arctic land masses together with the world's most extensive continental shelf, some of which might have been above sea level and ice-free during the last glaciation, have resulted in efficient dispersal of plants and animals so that many have circumpolar distributions and very large, extensive populations.

(d) Population dynamics

Life cycles of many plants and animals of the Arctic are cyclical. Lemming populations fluctuate in density on three- or four-year cycles. This cyclicity interacts with other trophic levels and results in similar population patterns for predators and for primary production which is reduced by grazing. Plants which require several years to flower, and in which flowering is followed by death of the flowering shoots, also follow cyclical patterns (Carlsson & Callaghan 1994). Invertebrates usually follow well determined life cycles with one generation per summer but opportunistic life cycles can occur although with the risk that non-resting stages of the life cycle, or non-hardened individuals may be killed by the early onset of winter (Webb *et al.* 1995). Population dynamics of migrant animals, mainly birds, depend on factors in areas other than the Arctic, such as hunting laws and changes in agriculture/land use along the migration routes and in their overwintering areas. Generally this decreases the population sizes of migrant birds, but in some cases population sizes can increase to the extent of increased grazing pressure denuding ecosystems (Jefferies *et al.* 1992).

4. Responses of the biota to environmental change

The fastest and most profound changes in Arctic biota are related to local land use and industrialization. Climate change impacts, in contrast, are likely to be more widespread and to occur over longer time scales.

(a) Responses of organisms

Environmental manipulation experiments simulating various aspects of climate change in different Arctic regions (Alaska, Swedish Lapland (figure 3e), Svalbard (figure 3a)) have shown that Arctic plant species can respond surprisingly quickly to increases in temperature and nutrient availability. Additions of nutrients have been given to plant communities to simulate the assumed increase in plant litter mineralization rates responding to higher temperatures and increasing deposition of atmospheric nitrogen. In general, nutrient addition treatments have great effects on plant performance, particularly in the most benign low altitude sub-Arctic environments (Havström *et al.* 1993). Increased temperatures stimulated the greatest plant responses at the most severe high Arctic sites (Havström *et al.* 1993; Wookey *et al.* 1993, 1995) where initial impacts of increased temperature were mostly restricted to plant development, reproduction and seed germinability (Wookey *et al.* 1993, 1995). Nichols (1995) found that spruce trees along a 1500 km transect through the Arctic tree line in eastern Canada produced pollen and cones in 1993 whereas no such production was seen 20 years before this.

Other experiments on Svalbard investigating responses of invertebrates to simulated climate change showed that, as in the plants, the response of an Arctic aphid was greatest where climate was harshest and response was apparent in the first year (Strathdee *et al.* 1994). In contrast, the responses of soil dwelling invertebrates were limited even after three years of environmental manipulation (Coulson *et al.* 1995), presumably because soils show less warming than the air (see below).

An 'individualistic' response of plants to the various environmental manipulation experiments resulted in a dampening of community productivity in Alaskan tussock tundra (Chapin & Shaver 1985) such that little variation in overall primary production occurred between treatments. However, the structure of the community changed over a decade such that graminoids showed a marked early response and then a dwarf shrub became dominant (Chapin *et al.* 1995). In the sub-Arctic, however, the compensatory mechanism of individualistic responses did not operate, and all higher plant species responded to environmental changes in the same direction (Jonasson 1992; Parsons *et al.* 1994). No species migrated into sub-Arctic birch heath experimental plots over the four years of the study.

In contrast, existing species can expand in response to environmental manipulations, particularly nutrient addition, in the most extreme environments where open ground is extensive (Fowbert & Lewis Smith 1994; Wyn Williams 1990; Wookey *et al.* 1995; figure 3a) and 'new' species can immigrate (C. H. Robinson, personal communication).

Data on direct responses of invertebrate and vertebrate animals to natural, rather than experimentally induced changes in temperature and moisture are rare. However, Järvinen (1994) found a significant positive correlation of increasing egg size in sub-Arctic birds, which increases survival, with temperature in the Finnish sub-Arctic and Tenow & Holmgren (1987) found a significant negative correlation between incidence of low winter temperature and insect damage during outbreaks of *Epirritia autumnata*, a moth caterpillar which defoliates sub-Arctic birch trees (figure 3f). In the last example, overwintering eggs were killed in cold depressions and in cold winters when temperatures dropped below $-36\,°C$. Any increase in temperature could lead to increased survival of eggs and greater destruction of birch forests.

(b) *Implications of climatic change for interactions between soil microbes and plants*

Any temperature increase will have greater effects on plant growth and the activity and development of the above-ground fauna than on the activity of soil fauna

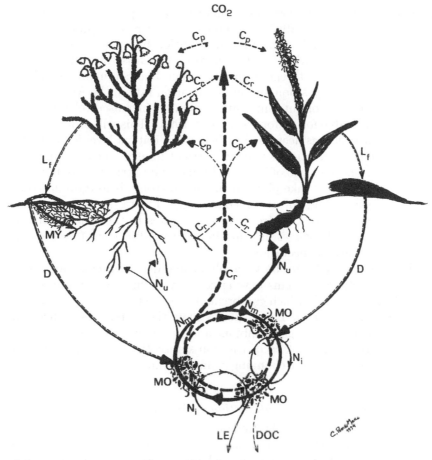

Figure 2. Schematic illustration of ecosystem carbon fluxes and nutrient circulation in tundra. Atmospheric CO_2 is fixed in the plants through photosynthesis (C_p). Dead parts of plants with organic carbon (carbon pathways with hatched arrows) and nutrients (pathways with continuous arrows) fall to the ground as litter (L_f) and go into the decomposition cycle (D) where they are transformed into soil organic matter. The organic matter with its nutrients undergo complicated transformations in the soil microbial biomass (MO) through cycles of mobilization/immobilization (N_m/N_i). Carbon is continuously lost to the atmosphere as CO_2 through microbial and plant respiration (C_r) and made available directly for plant photosynthesis or added to the atmospheric pool (CO_2). The nutrients are circulated within the microbial system of the soil or traded off and taken up by the plants (N_u). The plant species differ widely in nutrient uptake rates; here represented by *Cassiope tetragona* with slow uptake rate and *Polygonum viviparum* with high uptake rate. Some plants, e.g. *Cassiope tetragona* can circumvent the microbial mobilization/immobilization process by uptake of nutrients directly from organic matter decomposed by mycorrhizal fungi (MY) associated with their roots. Some carbon and nutrients can be lost from the system as dissolved organic matter (DOC) or nutrient leachate (LE) transported to adjacent systems. Climate change acts on the system mainly by controlling the rate of CO_2 uptake by plants and the activity of micro-organisms. Tundra ecosystems differ from others by having large stores of nutrients and carbon in the soil and microbial biomass and in having low rates of CO_2 uptake, i.e. low productivity.

and microbes, due to the insulation of the soil by plant cover and low conductivity of heat. Thus, below ground processes will be operating at lower temperature increases than those above ground. However, assuming that decomposition and mineralization rates increase after warming, fewer nutrients than anticipated may be available. Recent research (Jonasson *et al.* 1995) has shown that microbial uptake of nitrogen and phosphorus increases strongly after nutrient addition but shows little or no response after addition of a labile carbohydrate (sucrose) which is the response anticipated in most other ecosystems with energy-limited microbial communities. This suggests that any increased decomposition and mineralization rate in the Arctic may result in microbial immobilization of nutrients rather than an increased release rate of nutrients in the plant available form (Jonasson *et al.* 1993). This will proceed until the soil carbon to nutrient ratio has decreased to below the critical level for immobilization. As plant community production is usually nutrient limited, the assumption of significant soil warming and scenario of increased microbial nutrient immobilization imply that the main changes will take place in the heterotroph, belowground communities and not in the aboveground autotroph community. The result of these scenarios will be an increased emission of CO_2 from the ecosystems to the atmosphere (figure 2), but other environmental changes may act in opposite directions (see below) and soil warming may not match expectations.

In addition to the interactions between plants and decomposers, many plant species which dominate Arctic ecosystems have ericoid mycorrhizae or ectomycorrhizae that contain proteolytic enzymes which can break down complex organic, nitrogen-containing compounds (Read 1991). These species may, therefore, be supplied with nitrogen directly through their fungal symbiont and hence, circumvent the mobilization–immobilization cycle taking place in the non-symbiotic microbial community. How the mycorrhizae respond to a changing climate regime is however, poorly investigated.

(c) Implications for interactions between plants and animals

A more benign climate in the Arctic is likely to decrease mortality in most animal groups during summer which could increase herbivory. Increased herbivory is also likely to occur as a side-effect of increased plant growth, during which extra carbohydrates dilute the nutrient concentration in forage. In these circumstances, the herbivores must increase food intake to compensate for its lower nutrient concentration. Such a temperature dependent decline of forage quality has, indeed, been observed in plants during naturally occurring warm summers (Jonasson *et al.* 1986). Also higher levels of atmospheric carbon dioxide and UV-B radiation can decrease plant tissue 'quality' and adversely affect dependent organisms such as various invertebrate groups (Fajer *et al.* 1989; Couteaux *et al.* 1991) and fungal decomposers (Gehrke *et al.* 1995).

In contrast, winter mortality can increase in non-migrating mammals if winter temperatures – and particularly if the frequency of events with temperatures above zero – increases. Such thawing and freezing events could result in ice crust formation in the snow, and possibly formation of an ice-crust on the soil. This prevents larger mammals (reindeer, musk-oxen) from reaching the vegetation below the snowpack and small rodents living in the subnivean space from utilizing the vegetation encrusted in the ice.

(d) Changes in distributions of organisms and communities

The general stimulation of plant reproductive development by increased temperatures is likely to assume greatest significance in those areas with open, disturbed ground which can be colonized, e.g. high Arctic deserts and polar semi-deserts (figure 3*a*), Arctic fellfields, thermokarst landscapes (figure 3*b*), and areas disturbed by man around settlements, industries, mines, oil wells, roads, etc. (Forbes 1995).

Tundra ecosystems are characterized by a mosaic of habitats with different communities, e.g. polygonal tundra with moist, low-centred polygons, dry polygon ridges and wet polygon troughs (figure 3*c*). Latitudinal distances between major geographical vegetation zones on the other hand tend to be large (hundreds of km). Thus the slow migration rates of species along extensive latitudinal gradients is likely to result in only longer term changes in communities whereas the colonization of changing habitats by species from neighbouring habitats is likely to be a faster process. However, exceptions may occur in those areas acting as 'refugia' for species with a southerly distribution, e.g. pockets of trees in the tundra (Landhausser & Wein 1994) and erect shrubs along river valleys in mid Arctic tundras. Such refugia with local favourable microclimates might provide 'innocula' for the invasion of extensive areas when the general climate changes. In the high Arctic (e.g., figure 3*a*), the initial changes in communities might be more subtle as ecotypes of species such as *Saxifraga oppositifolia* from microclimatically favourable microhabitats displace those from less favourable sites during climate warming (Crawford *et al.* 1993). Other exceptions to the expected gradual change in distribution of organisms occur where corridors for migration are created by human activities (Forbes 1995).

Animals, with their greater mobility than plants, can change their distributions rapidly and this phenomenon has been used to infer rapid changes in climate during the Holocene from changes in distributions of *Diptera* (Coope 1975).

The mechanisms for change in plant community structure are likely to be competitive exclusion of northern species by southern species or faster growing neighbouring species and ecotypes (Callaghan & Jonasson 1995). This has been inferred from the sensitivity of many Arctic plant species to shade yet their general ability to grow in more southerly latitudes when competitors are artificially excluded. Some Arctic species, particularly those with large below-ground biomass, may however, suffer an adverse carbon balance in warmer climates.

Assuming a doubling of atmospheric CO_2, and no constraints on the migration of species, Emanuel *et al.* (1985) and Leemans (1989) calculated that tundra areas would decrease in extent by between 20 and 32% because of the northwards expansion of the boreal forest biome (figure 3*e*). However, tree migration rates are slower than predicted rates of climate shift. For example, a warming of 2 °C could result in a 4–5° latitude northward shift of the climate zone currently associated with the taiga of Eurasia (Velichko *et al.* 1990), i.e. a shift of 400–500 km by the year 2020. If the taiga could migrate at the same rate, tundra would be totally displaced from the Eurasian mainland by 2020. However, the migration rates of taiga trees are only about 10–300 m a^{-1} for conifers (Nichols 1967; Chertovskij *et al.* 1987 quoted in Razzhivin 1995) to 130 to 1000 m a^{-1} for deciduous alder and birch (Velichko *et al.* 1990; Chertovskij *et al.* 1987 quoted in Razzhivin 1995).

The inequality in rate of climate and vegetation shift will subject large areas of vegetation to supra-optimal climate regimes where damage from extreme weather, fire, pests etc will increase. It is likely that the individualistic responses of plant species to climate change will result in the disintegration of current communities and the formation of new assemblages of plants.

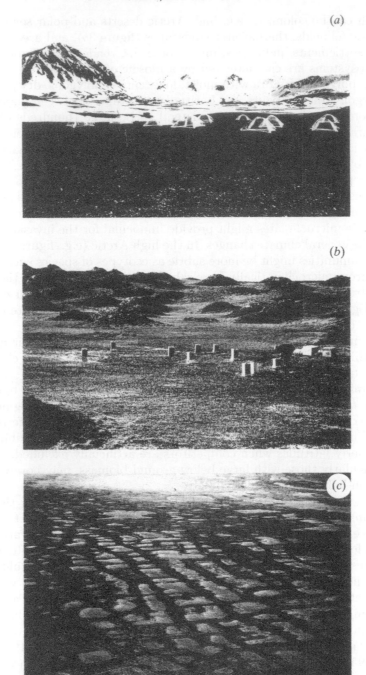

Figure 3. Arctic landscape types. (*a*) High Arctic polar semi-desert near Ny Ålesund, Svalbard showing temperature manipulation experiments on scattered *Dryas octopetala* vegetation. Photo credit: T. V. Callaghan. (*b*) High Arctic vegetation and thermokarst scenery dominated by 'baidgerakhs', i.e. raised polygon centres accentuated by rapidly thawing polygon troughs. Ostrov Faddeyevskiy, New Siberian Islands. Measurements of methane fluxes are in progress. Photo credit: S. E. Jonasson. (*c*) Polygonal/tetragonal coastal tundra. Olenekskiy Bay, North-central Siberia, showing sunken and flooded polygon centres, raised polygon rims and polygon troughs. Photo credit: T. V. Callaghan.

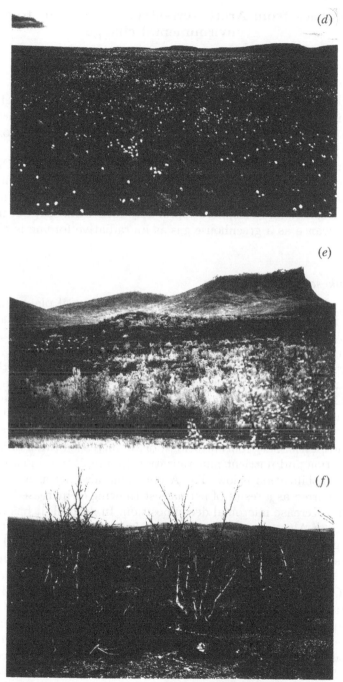

Figure 3. (*d*) Mid-Arctic tussock tundra dominated by *Eriophorum vaginatum* near the Yana Delta, east of the Taimyr Peninsula, North Siberia. Photo credit: S. E. Jonasson. (*e*) Sub-Arctic tundra/taiga ecotone dominated by mountain birch trees (*Betula* pubescence ssp *tortuosa*) and dwarf shrub heath vegetation, near Abisko, Swedish Lapland. A randomized block temperature enhancement experiment can be seen beyond the tree line. Photo credit: J. Busch. (*f*) Sub-Arctic mountain birch forest destroyed by the defoliating moth caterpillar *Epirrita autumnata* near Kevo, Finland. Pest outbreaks are currently controlled by low winter temperatures which kill eggs. Photo credit: T. V. Callaghan.

5. Feedback from Arctic terrestrial ecosystems to global environmental change

(a) Carbon cycling

(i) Carbon stocks, trace gas fluxes

The large stock of carbon stored in the tundra soils, estimated to be about 11% of that stored in soils globally (Schlesinger 1984), is a potential important source of greenhouse gases (CO_2 and CH_4) to the global atmosphere. Carbon dioxide is formed by autotrophic and heterotrophic respiration and is consumed and converted into organic carbon by autotrophic photosynthesis (figure 2). Methane (CH_4) is formed by bacterial degradation of organic matter in anaerobic environments and can be consumed by aerobic bacteria. Its direct contribution to changes in carbon pool sizes in the atmosphere and biosphere is probably rather limited, but methane has a particular significance as a greenhouse gas as its radiative forcing is about 30 times that of CO_2.

(ii) Carbon sink–source relationships

The tundra will act as an atmospheric source of carbon if plant and soil respiratory processes release more carbon per unit time than that fixed in plants (figure 2). This will occur if below-ground processes are stimulated more by changes in temperature and/or moisture than plant productivity. Such a scenario is possible in systems where low nutrient availability limits both plant and microbial growth (see above). The temperature-stimulated turnover of soil organic matter would lead to increased emission of CO_2 at the same time as the microorganisms immobilize the nutrients and prevent any substantial increase of plant production. On the other hand, the Arctic can act as a sink for atmospheric carbon if plant productivity is enhanced by increasing atmospheric CO_2 concentrations, or by increased availability of nutrients from decomposition and nutrient mineralization (figure 2), i.e. in cases when microbial nutrient immobilization is low. The Arctic could also act as a sink for carbon if waterlogging occurred as a result of permafrost thawing or increased precipitation as these conditions decrease microbial decomposition. In general, it has been suggested that the tundra of Alaska acted as a net sink for carbon throughout the Holocene by sequestering atmospheric carbon in organic soils and peat (Oechel et al. 1993).

Experiments by Billings et al. (1982; 1983) on soil microcosms from Alaska show that a drying and warming of tundra soils will lead to the increased emission of carbon. A 4 °C temperature increase could increase loss of carbon by 60–80 g m^{-2} if the water table is 5 cm below the surface and by 130–160 g m^{-2} with an 8 °C temperature increase if the water table is at a depth of 10 cm. An increase of 4 °C could therefore, increase decomposition rates and the evolution of soil carbon by 1 Pg in tundra areas and 0.5 to 2 Pg in boreal areas (Lashof 1989). In wet tundra and boreal areas which currently emit about 40 Tg methane per year to the atmosphere, a 4 °C increase in mean annual temperature could increase methane evolution by 45–65% (Melillo et al. 1990).

Due to the large regional and local variations over the Arctic, it seems realistic to assume that the sink–source relationship will probably vary on both a local basis (e.g. plant community, ecosystem or landscape) and regionally across the latitudinal extent of the Arctic. This has indeed been suggested for Alaskan tundra. Oechel et al. (1993) reported a recent change from sink (see above) to source status of the

Alaskan tundra, particularly the wet coastal tundra in comparison to the inland mesic tussock tundra.

Given the predictions of increased plant productivity in the Arctic if the climate warms (Melillo *et al.* 1994), and the proportionately low carbon stores in the high Arctic, the high Arctic will probably become a net sink for atmospheric CO_2. In contrast, the tundra in the middle Arctic with a closed vegetation cover will probably increase its sink action marginally, or possibly act as a source of CO_2 due to increased microbial respiration but little increase in productivity. The southern dwarf shrub and tall shrub tundra, bordering the boreal forest may, again, act as sinks because they have a high potential for carbon fixation even under nutrient limited conditions due to a high production of woody tissues with low nutrient content.

Production and emission of methane, which is a second process by which the soil and the atmosphere exchange carbon, occur principally in water-saturated areas of the tundra, for instance in the wet sedge tundra bordering the northern coasts of Siberia and Alaska. Methane is, on the other hand consumed by CH_4-oxidising bacteria in drier tundra areas. Both the quantity of methane produced over the tundra areas and the balance between methane production and consumption is poorly known. However, measurements of methane fluxes at regular intervals along the coast of the Siberian tundra (Christensen *et al.* 1995) have indicated that net emission is generally restricted to entirely water-saturated soils, e.g. low centres of tundra polygons (figure 3c). Mesic sites, e.g. rims of the same polygons (some 15–20 cm higher) generally showed no net emission and could even consume methane.

(b) Albedo

The Arctic currently cools the earth by reflecting more incoming solar radiation than it absorbs. Ice and snow have the greatest reflectance and tundra vegetation has higher reflectivity (albedo) than dark, coniferous boreal forests. A result of intensive warming at high latitudes will be an eventual decrease in the extent of snow and ice on land and a northward shift in the boreal forest zone. This positive albedo feedback will be greater than that of the emissions of carbon from the biosphere and will increase the inequality of warming between the poles (mainly the Arctic) and the equator (Lashof 1989).

In conclusion, responses of individual plant species to single environmental factors are relatively well known, but impacts of environmental change on Arctic animals are less well known. At the ecosystem level, the impacts are particularly complex and difficult to foresee. The main reasons for this are the many interactions which exist within ecosystems and between several concurrently changing environmental factors.

We are indebted to the Swedish Polar Secretariat and the Swedish Royal Academy of Sciences for enabling our participation in the Russian Swedish Tundra 94 Expedition which broadened the representativeness of this contribution. We are also grateful to the Danish Natural Environment Research Council, the Swedish Environmental Protection Agency and the UK NERC Arctic Terrestrial Ecology Special Topic Programme for grants enabling our participation in Arctic ecological research. It is a pleasure to thank all of our co-workers for their collaboration and we would like to thank Christina Pomar for drawing figure 2 and Jonathan Kirkham for preparing figure 1.

References

Billings, W. D. 1992. Phytogeographic and evolutionary potential of the Arctic flora. In Chapin *et al.* (1992), pp. 91–109.

Billings, W. D., Luken, J. O., Mortensen, D. A. & Peterson, K. M. 1982 Arctic tundra: a source or sink for atmospheric carbon dioxide in a changing environment? *Oecologia (Berl)* **53**, 7–11.

Billings, W.D., Luken, J. O., Mortensen, D. A. & Peterson, K. M. 1983 Increasing atmospheric carbon dioxide: Possible effects on Arctic tundra. *Oecologia Berl.* **58**, 286–289.

Bryant, J. P. & Reichardt, P. B. 1992 Controls over secondary metabolite production by Arctic woody plants In Chapin *et al.* 1992: pp. 377–390.

Callaghan, T. V., Sonesson, M. & Sømme, L. 1992 Responses of terrestrial plants and invertebrates to environmental change at high latitudes. *Phil. Trans. R. Soc. Lond.* B **338**, 279–288.

Callaghan, T. V. & Jonasson, S. 1995 Implications for changes in Arctic plant biodiverity from environmental manipulation experiments. In *Arctic and alpine biodiversity: patterns, causes and ecosystem consequences* (ed. F. S. Chapin III & C. Körner), pp. 149–164. Heidelberg: Springer Verlag.

Callaghan, T. V., Oechel, W. C., Gilmanov, T., Holten, J. I., Maxwell, B., Molau, U., Sveinbjörnsson, B. & Tyson, M. (eds) 1995 Global change and Arctic terrestrial ecosytems. *Proc. Int. Conf., Oppdal, Norway, 21–26 August 1993.* Brussels: Commission of the European Communities Ecosystems Research Report. (In the press.)

Carlsson, B. Å. & Callaghan, T. V. 1994 Impact of climate change factors on *Carex bigelowii*: implications for population growth and spread. *Ecography* **17**, 321–330.

Chapin, F. S. III & Shaver, G. R. 1985 Individualistic response of tundra plant species to environmental manipulations in the field. *Ecology* **66**, 564–576.

Chapin, F. S. III, Jefferies, R. L., Reynolds, J. F., Shaver, G. R. & Svoboda, J. 1992 *Arctic ecosystems in a changing climate: an ecological perspective.* San Diego: Academic Press.

Chapin, F. S. III, Shaver, G. R., Giblin, A. E., Nadelhoffer, K. G. & Laundre, J. A. 1995 Responses of Arctic tundra to experimental and observed changes in climate. *Ecology* **76**, 694–711.

Christensen, T. R., Jonasson, S., Callaghan, T. V. & Havström, M. 1995 Spatial variation in high-latitude methane flux along a transect across Siberian and European tundra environments. *J. geophys. Res.* (In the press.)

Coope, G. R. 1975 Climatic fluctuations in Northwest Europe since the last interglacial, indicated by fossil assemblages of Coleoptera. In *Ice ages: ancient and modern* 1975 (ed. A. E. Wright & F. Moseley), pp. 153–168. Liverpool: Seel House Press.

Coulson, S., Hodkinson, I. D., Webb, N. R., Block, W., Worland, M. R., Bale, J. S., Strathdee, A. T. & Wooley, C. 1995 Effects of experimental temperature elevation on high Arctic soil microarthropod populations. *Polar Biol.* (In the press.)

Couteaux, M. M., Mousseau, M., Celerier, M. L. & Bottner, P. 1991 Increased atmospheric CO_2 and litter quality: decomposition of sweet chestnut leaf litter with animal food webs of different complexities. *Oikos* **61**, 54–64.

Crawford, R. M. M., Chapman, H.M., Abbott, R.J. & Balfour, J. 1993 Potential impact of climatic warming on Arctic vegetation. *Flora* **188**, 367–381.

Crawford, R. M. M., Chapman, H. M. & Hodge, H. 1994 Anoxia tolerance in high Arctic vegetation *Arctic alpine Res.* **26**, 308–312.

Dickinson, R. E. & Hanson B. 1984 Vegetation-albedo feedbacks. In *Climate processes and climate sensitivity* (ed. J. Hansen & T. Takahashi), pp. 180–186, Geophysical Monograph 29. Washington, DC: American Geophysical Union.

Emanuel, W. H., Shugart, H. H. & Stevenson, M. P. 1985 Climate change and the broad-scale distribution of terrestrial ecosystem complexes. *Climate Change* **7**, 29–43.

Fajer, E. D., Bowers, M. D. & Bazzaz, F. A. 1989 The effects of enriched carbon dioxide atmospheres on plant–insect herbivore interactions. *Science, Wash.* **243**, 1198–1200.

Forbes, B. C. 1995 Effects of surface disturbance on the movement of native and exotic plants under a changing climate. In Callaghan *et al.* (1995).

Fowbert, J. A. & Lewis Smith, R. I. L. 1994 Rapid population increases in native vascular plants in the Argentine Islands, Antarctic Peninsula. *Arctic alpine Res.* **26**, 290–296.

Gehrke, C., Johanson, U., Callaghan, T. V., Chadwick, D. & Robinson, C. H. 1995 The impact of enhanced ultaviolet B-radiation on litter quality and decomposition processes in *Vaccinium* leaves from the sub-Arctic. *Oikos* **72**, 213–222.

Goreau, T. J. 1990 Balancing atmospheric carbon dioxide. *Ambio* **19**(5), 230–236.

Havström, M., Callaghan, T. V. & Jonasson, S. 1993 Differential growth responses of *Cassiope tetragona*, an Arctic dwarf shrub, to environmental perturbations among three contrasting high- and sub-Arctic sites. *Oikos* **66**, 389–402.

Havström, M., Callaghan, T. V., Jonasson, S. & Svoboda, J. 1995 Little Ice Age temperature estimated by growth and flowering differences between subfossil and extant shoots of *Cassiope tetragona*, an Arctic heather. *Funct. Ecol.* **9**. (In the press.)

Hendry, G. A. F. 1993 Forecasting the impact of climatic change on natural vegetation with particular reference to boreal and sub-boreal floras. In *Impacts of climatic change on natural ecosystems with emphasis on boreal and Arctic/alpine areas*. 1993 (ed. J. I. Holten, G. Paulsen & W. C. Oechel), pp. 136–150. Trondheim, Norway: Norwegian Institute for Nature Research (NINA) and the Directorate for Nature Management (DN).

Järvinen, A. 1994 Global warming and egg size of birds. *Ecography* **17**, 108–110.

Jefferies, R. L., Svoboda, J., Henry, G. H. R., Raillard, M. & Ruess, R. 1992 Tundra grazing systems and climatic change. In Chapin *et al.* (1992), pp. 391–412.

Jonasson, S. 1992 Plant responses to fertilization and species removal in tundra related to community structure and clonality. *Oikos* **63**, 420–429.

Jonasson, S., Bryant, J. P., Chapin, F. S. III & Andersson, M. 1986 Plant phenols and nutrients in relation to variations in climate and rodent grazing. *Am. Naturalist* **128**, 394–408.

Jonasson, S., Havström, M., Jensen, M. & Callaghan, T. V. 1993 In situ mineralization of nitrogen and phosphorus of Arctic soils after perturbations simulating climate change. *Oecologia, Berl.* **95**, 179–186.

Jonasson, S., Michelsen, A., Schmidt, I. K., Nielsen, E. V. & Callaghan, T. V. 1995 Uptake of N and P by microbes in two Arctic soils after addition of nutrients and sugar: implications for plant nutrient uptake. *Oecologia, Berl.* (Submitted.)

Jones, P. D. & Briffa, K. R. 1992 Global surface air temperature variations during the twentieth century, 1. Spatial, temporal and seasonal detail. *The Holocene* **2**, 165–179.

Jonsdottir, I. S. & Callaghan, T. V. 1990 Intraclonal translocation of ammonium and nitrate nitrogen in *Carex bigelowii* using ^{15}N and nitrate reductase assays. *New Phytol.* **114**, 419–428.

Landhauser, S. M. & Wein, R. W. 1994 Postfire vegetation recovery and tree establishment at the Arctic treeline: climate-change–vegetation-response hypotheses. *J. Ecol.* **81**, 665–672.

Lashof, D. A. 1989 The dynamic greenhouse: feedback processes that may influence future concentrations of atmospheric trace gases and climatic change. *Climate Change* **14**, 213–242.

Leemans, R. 1989 Possible changes in natural vegetation patterns due to global warming. In *Der Treibhauseffect: das Problem – mögliche Folgen – erforderliche Massnahmen* 1989 (ed. Hackel), pp. 105–121. Luxemburg: Akademie für Umwelt und Energie.

Melillo, J. M., Callaghan, T. V., Woodward, F. I., Salati, E. & Sinha, S. K. 1990 Effects on ecosystems. In *Climate Change, the IPCC Scientific Assessment*. 1990 (ed. J. T. Houghton, G. J. Jenkins & J. J. Ephraums), pp. 282–310. Cambridge University Press.

Melillo, J. M., McGuire, A. D., Kicklighter, D. W., Moore, B. III, Vorosmarty, C. J. & Schloss, A. L. 1993 Global climate change and terrestrial net primary production. *Nature, Lond.* **363**, 234–240.

Mitchell, J. F. B., Manabe, S., Tokioka, T. & Meleshko, V. 1990 Equilibrium climate change. In Houghton *et al.* (1990), pp. 131–172.

Mølgaard, P. 1982 Temperature observations in high Arctic plants in relation to microclimate in the vegetation of Peary Land, North Greenland. *Arctic Alpine Res.* **14**, 105–115.

Nichols, H. 1967 The post-glacial history of vegetation and climate at Ennadai Lake, Keewatin, and Lynn Lake, Maitoba (Canada). *Eiszeitaler Gegenwart* **18**, 176–197.

Nichols, H. 1995 Reproductive changes in the Canadian Arctic tree-line: possible greenhouse effect? *Nature, Lond.* (Submitted.)

Oechel, W.C., Hastings, S.J., Jenkins, M., Reichers, G., Grulke, N. & Vorlitis, G. 1993 Recent change of Arctic tundra ecosystems from a carbon sink to a source. *Nature* **361**, 520–526.

Parsons, A. N., Welker, J. M., Wookey, P. A., Press, M. C., Callaghan, T. V. & Lee, J. A. 1994. Growth responses of four sub-Arctic dwarf shrubs to simulated environmental change. *J. Ecol.* **82**, 307–318.

Razzhivin, V. Yu. 1995 Effects of climate change and dynamics of tundra plant communities in far eastern Asia. In Callaghan *et al.* (1995). (In the press.)

Read, D. J. 1991 Mycorrhizas in ecosystems – Nature's response to the 'law of minimum'. In *Frontiers in mycology* (ed. D. L. Hawksworth), pp. 101–130. Regensburg: Lectures 4th Int. mycolog. Congress, 1990.

Schlesinger, W. H. 1984 Soil organic matter: a source of atmospheric CO_2. In *The role of terrestrial vegetation in the global carbon cycle, methods of appraising changes* (ed. G. M. Woodwell), SCOPE 23, pp. 111–127. Chichester: John Wiley.

Shaver, G.R. & Chapin III, F.S. 1980 Response to fertilization by various plant growth forms in an Alaskan tundra: nutrient accumulation and growth. *Ecology* **61**, 662–675.

Strathdee, A. T., Bale, J. S., Strathdee, F. C., Block, W. C., Coulson, S. J., Webb, N. R. & Hodkinson, I. D. 1995 Climatic severity and the response to temperature elevation of Arctic aphids. *Global Change Biol.* **1**, 23–28.

Tenow, O. & Holmgren, B. 1987 Low winter temperatures and an outbreak of *Epirrita autumnata* along a valley of Finnmarksvidda, the 'cold-pole' of northern Fennoscandia. In *Climatological extremes in the mountain, physical background, geomorphological and ecological consequences* (ed. H. Axelsson & B. Holmgren), pp. 203–216. Uppsala, Sweden: Dept. of Physical Geography of the University of Uppsala, UNGI Rapport 15.

Tissue, D. T. & Oechel, W. C. 1987 Response of *Eriophorum vaginatum* to elevated CO_2 and temperature in the Alaskan tussock tundra. *Ecology* **68**, 401–410.

Velichko, A. A., Borisova, O. K., Zelikson, E. M. & Nechaev, V. P. 1990 An assessment of dynamics of natural geosystems in the forest and tundra zones under anthropogenic climatic change. Quoted in *Climate change: the IPCC impacts assessment* (ed. W. J. McG. Tegart, G. W. Sheldon & D. C. Griffiths), ch. 3. Canberra: Australian Government Publishing Service.

Webb, N. R, Hodkinson, I. D., Coulson, S., Bale, J. S., Strathdee, A. T. & Block, W. 1995 Life history and ecophysiological responses to temperature in Arctic terrestrial invertebrates. In Callaghan *et al.* (1995). (In the press.)

Wookey, P. A., Parsons, A. N., Welker, J. M., Potter, J. A., Callaghan, T. V., Lee, J. A. & Press, M. C. 1993 Comparative responses of phenology and reproductive development to simulated climate change in sub-Arctic and high Arctic plants. *Oikos* **67**, 490–502.

Wookey, P. A., Robinson, C. H., Parsons, A. N., Welker, J. M., Press, M. C., Callaghan, T. V. & Lee, J. A. 1995 Environmental constraints on the growth, photosynthesis and reproductive development of *Drays octopetala* at a high Arctic polar semi-desert, Svalbard. *Oecologia, Berl.* (In the press.)

Wyn Williams, D. 1990 Microbial colonization processes in Antarctic fellfield soils – an experimental overview. *Polar Biol.* **3**, 164–178.

Discussion

H. NICHOLS (*Department of EPO Biology, University of Colorado, USA*). Previous commentary of mine on the sensitivity of the Arctic tree-line to climatic change referred to palaeoclimatic data and meteorological records to argue that this vegetation boundary has been one of the most responsive ecotones to register environmental change. Palynological evidence demonstrates mid-Holocene (hypsithermal) movements of the Canadian Arctic tree-line of up to 400 km into the tundra, in response to a natural (Milankovitch) warming cycles, locally of about +4 °C mean July temperatures (i.e. +1 °C mean = 100 km advance of tree-line). Estimates of woodland advances into Arctic tundra are approximately 300 m a^{-1} under climatic warming comparable to a full-scale $2 \times CO_2$ greenhouse effect (about 3–4 °C increased mean summer temperatures). Since the global warming predicted from the anthropogenic greenhouse effect includes a strong polar warming scenario, we have sought to re-visit the sites of Nichols's 20-year old expeditions to record pollen and cone production at the boreal forest-tundra ecotone. In 1972 and 1973 a transect of pollen sampling sites was studied, from the northern woodland edge and into the Keewatin tundra, passing through clumps of mature dwarf spruce trees. At that time trees were barren, with no consistent pollen release or cone formation, indicative of then prevailing summer climatic limitations. In June and July 1993 several of the 1972–73 sites plus additional locations were visited, forming an east–west transect of about 1500 km in the Northwest Territories of Canada, from inland (west) of Hudson bay to an area north of the Great Bear Lake, south of the Arctic Ocean coastline. Eight locations were spaced along this transect, and in late June-early July Nichols set out pollen traps at each site to record regional pollen fallout from the forest to the south as well as that from the isolated groups of spruce trees. This design mimicked the 1972–73 project, the aim being to distinguish 'spikes' of local spruce pollen production from the isolated tree groups, from the background forest pollen deposition.

This east–west transect of sampling sites was repeated later in July, and as a result spruce pollen release was observed at a number of sites which did not produce pollen in 1972 and 1973. Even more impressive was the observation that at all the sites, from the whole length of the transect of 1500 km, cones were being formed on spruce trees where 20 years ago the trees were barren. At sites 150 miles (250 km) from the woodland edge, out into the tundra, dwarf spruce less than one metre high were bearing this year's (1993) cones, along with those from previous years. There is no proof that this is due to anthropogenic warming (it could be a natural 'flickering' of climate), but it is a phenomenon which would have to be observed if polar warming due to the greenhouse effect were to be credible. It is important in this respect to continue these observations to identify a trend of continuing cone production versus a return to barrenness on the part of these marginal trees.

T. V. CALLAGHAN. Prof. Nichols has described an interesting study. The isolated trees beyond the latitudinal treeline could play a particularly important role in providing foci for the expansion of the boreal forest in a warmer climate. I agree that more frequent observations are necessary, particularly as what you have described cannot, on the basis of two observations periods, be separated from the innate cyclicity in biological phenomena in the Arctic which we have described in our paper.

R. B. HEYWOOD (*British Antarctic Survey, UK*). Could temperate (plant) species have considerably faster reproductive processes than contemporary Arctic species. Has Professor Callaghan considered the effect on community structure of the former 'leap-frogging', as it were, the latter during colonization of regions made less hostile by the effects of global warming.

T. V. CALLAGHAN. The process of temperate plants 'leap-frogging' into Arctic plant communities during climate warming usually depends on the availability of sites for establishment and efficient dispersal, assuming that climatic conditions are appropriate for subsequent survival. Available sites and dispersal are associated in particular with human disturbance. Thus, the greatest probability of observing this process occurs where roads and pipelines have been constructed and industrialization has disturbed or destroyed natural Arctic plant communities. Under natural conditions, the vast latitudinal distances separating temperate plants form the Arctic and the dominance of closed vegetation in the more southerly regions of the Arctic, together constrain the process of 'leap-frogging'.

R. B. HEYWOOD. The influence of isolation cannot be ignored in any discussion in the level of endemism in polar marine ecosystems. Most marine organisms are shallow water species. The Arctic Ocean has vast areas of shallow coastal seas which are continuous with the coastal waters of all the Northern Hemisphere continents. There is no impedance to migration in and out of the Arctic Ocean. In contrast the narrow coastal waters of Antarctica are separated from those of the Southern Hemisphere continents by the vast distances of the very deep Southern Ocean.

T. V. CALLAGHAN. Dr Heywood has underlined the essential role for the geography of the Arctic in controlling endemism and trends in biodiversity in general, and that this is equally important in marine and terrestrial ecosystems.

P. A. WOOKEY (*Department of Biological Sciences, University of Exeter, UK*). Professor Callaghan emphasizes early in his paper the large storage of global soil carbon in tundra ecosystems (around 11–14% of the total) and the potential for release of substantial quantities of this to the atmosphere as a result of soil warming and accelerated decomposition processes. This could indeed act as a positive feedback to climate change if assimilation of atmospheric CO_2 by plants is less than that released by heterotrophic respiration processes in the decomposition subsystem. The likelihood of such a large net efflux of CO_2 from tundra ecosystems to the atmosphere may, however, be called into question in view of the palaeoecological record of rates of carbon accumulation during the Holocene climatic optimum (6900–4800 BP): these records suggest that carbon accumulation rates in Arctic Alaskan ecosystems were highest when the climate was some 2–4 °C warmer than at present (see Marion & Oechel 1993). What is the likely explanation of this apparent paradox?

T. V. CALLAGHAN. As you noted, gaseous carbon flux from the biosphere to the atmosphere and the converse organic carbon storage in soils, result from the balance between primary production and decomposition. It is possible to envisage increased rates of primary production responding to higher temperatures during the Holocene climatic optimum but decreased rates of decomposition responding to cooler soils insulated by higher biomass, or wetter soils if the climatic optimum was also a wetter period. More problematic, however, is how increased primary production could be sustained with reduced nutrient availability associated with slower decomposition.

Additional references

Marion, G. M. & Oechel, W. C. 1993 Mid- to late-Holocene carbon balance in Arctic Alaska an its implications for future global warming. *The Holocene* **3**, 193–200.

Climate change and biological oceanography of the Arctic Ocean

By R. Gradinger

*Institut für Polarökologie der Christian-Albrechts-Universität Kiel,
Wischhofstr. 1-3, Geb. 12, D-24148 Kiel, Germany*

Polar environments are characterized by unique physical and chemical conditions for the development of life. Low temperatures and the seasonality of light create one of the most extreme habitats on Earth. The Arctic sea ice cover not only acts as an insulator for heat and energy exchange processes between ocean and atmosphere but also serves as a unique habitat for a specialized community of organisms, consisting of bacteria, algae, protozoa and metazoa. The primary production of sea ice algae may play a crucial role in the life cycle of planktonic and benthic organisms. Thus, a reduction of the sea ice extent due to environmental changes will influence the structure and processes of communities living inside the ice and pelagic realms.

---

1. Introduction

The Arctic Ocean is an enclosed sea area with two major connections to the surrounding seas, the shallow Bering Sea and the relatively deep Fram Strait. The Arctic marine environment is in its present state one of the most extreme habitats on Earth. Strong seasonal variations of some parameters, such as solar radiation, are in contrast to the relative stability of others, such as water temperature. Organisms living in polar oceans are well adapted to these environmental conditions. Climatic changes will therefore influence the structure of the Arctic marine communities, as already indicated by the geological record: the Pleistocene warming at approximately 1.5 Ma coincides with a drastic increase in North Atlantic species and calcareous organisms in the sediment record as a result of changes in the water mass exchange between the Arctic Ocean and its surrounding seas (Clarck 1990).

The emission of trace gases to the atmosphere by anthropogenic activities may lead to similar changes, but on much shorter time scales. Global models studying the effect of CO_2 increase in the Earth's atmosphere showed the largest temperature increase in the Arctic (Mitchell *et al.* 1990). Recent measurements from Alaska already demonstrate a temperature increase of approximately 1.5 °C during the past decade (Oechel & Vourlitis 1994).

An increase in atmospheric, and consequently sea surface, temperature will have a high impact on the Arctic marine epipelagic system. Based on present knowledge of the Arctic marine pelagic system, possible changes will be discussed in this contribution, but the developed scenario will be in no sense predictive. It is obvious that the response of the Arctic marine ecosystem will largely depend on the fate of its most characteristic realm, the permanent sea ice cover.

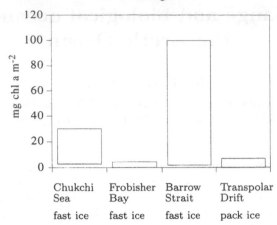

Figure 1. Algal biomass in Arctic sea ice. Data for Chukchi Sea from Clasby *et al.* (1973); Frobisher Bay from Grainger (1979); Barrow Strait from Smith *et al.* (1989) and Transpolar Drift from Gradinger (unpublished data).

2. The sea ice realm

Sea ice covers between 7 and 14×10^6 km^2 of the Arctic Ocean (Maykut 1985). Its existence largely influences the material and energy exchange between ocean and atmosphere and is therefore a crucial parameter in the modelling of environmental changes in polar areas. In contrast to Antarctica, about 50% of the Arctic sea ice floes survive summer melting and thus reach thicknesses of more than 2 m (for a detailed comparison between Arctic and Antarctic sea ice properties, see Spindler 1990).

Sea ice consists of a mixture of ice crystals and brine channels, which form a three-dimensional network of tubes and channels with typical diameters of 200 μm (Weissenberger *et al.* 1992) within the ice matrix. The brine salinity and the total volume of the brine channels as percentage of the ice volume are dependent on ice temperature and total salt content. For example, a decrease in the ice temperature from -4 °C to -10 °C leads to growth of ice crystals and thus an increase in brine salinity from 70 to 144 psu (Assur 1958), as well as a decrease in the brine volume.

Despite these harsh environmental conditions, a specialized community has developed and adapted to live within the brine channel system. Diatoms are the dominant primary producers and may contribute more than 90% of the total algal biomass (Poulin 1990). The seasonal development of the sea ice algae is mainly controlled by abiotic parameters. The onset of algal growth in spring is triggered by an increase in available light intensities after the dark polar winter. Sea ice, and especially its snow cover, reduces the incoming radiation by more than 90% due to high albedo. Therefore, ice algae are already adapted to start growing under extremely low light intensities (2–10 μmol m^{-2} s^{-1}; Horner & Schrader 1982). The biomass built up by sea ice algae during the Arctic summer varies between 1 and 100 mg Chl a m^{-2} (figure 1). Highest concentrations have been observed in fast ice areas of the Canadian shelf (Clasby *et al.* 1973; Smith *et al.* 1989), while the concentrations within the multiyear ice floes of the transpolar drift system are one or two orders of magnitude lower (Gradinger, unpublished data).

Large fluctuations in temperature, and therefore brine salinity, restrict life within the Arctic ice floes to the lowermost decimetres, and so-called sea ice bottom com-

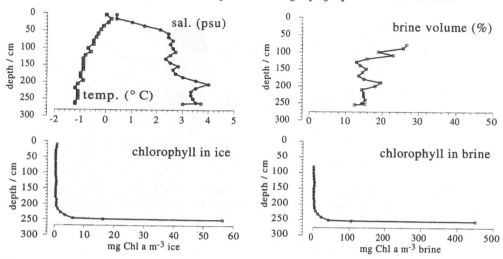

Figure 2. Temperature, salinity, brine volume and algal biomass in an Arctic multiyear ice floe
(Gradinger, unpublished data).

munities are formed (Horner 1985). Figure 2 shows an example of the chlorophyll
distribution in an Arctic multiyear ice floe, sampled in the East Greenland Current
in August 1994 (Gradinger, unpublished data). Low salinities and relative high temperatures are idiosyncratic for Arctic summer sea ice. The calculated brine volume
based on the equations by Frankenstein & Garner (1967) varies between 10 and
30% of the total ice volume. The chlorophyll profile clearly shows a well developed
bottom community with concentrations above 50 mg Chl a m^{-3} ice in the lowermost
centimetres. The actual algal concentration within the brine channel system is even
higher exceeding values of 400 mg Chl a m^{-3} brine. This high algal biomass serves as
the food source for a variety of proto- and metazoans (figure 3), which are mostly
smaller than 1 mm. In shallow sea areas, nematoda and crustaceans are the dominating organism groups (Carey & Montagna 1982; Cross 1982; Kern and Carey 1983;
Grainger *et al.* 1985), while a distinct community inhabits multiyear ice floes, with
ciliates and turbellarians as most abundant taxa (Gradinger *et al.* 1991).

The high algal biomass inside Arctic ice floes is used by pelagic and benthic organisms during parts of their life cycle. Carey & Montagna (1982) observed larvae of
benthic polychaetes and molluscs inside Arctic sea ice, and Kurbjeweit *et al.* (1993)
made a similar observation for the Antarctic pelagic copepod, *Stephos longipes*. For
these organisms, ice floes serve as a kind of 'kindergarten' to the juveniles, providing
both food and shelter against possible predators.

3. The under-ice realm

The boundary-layer between Arctic ice floes and the water column forms the
habitat for a specific community of organisms. Diatoms, mainly the species *Melosira
arctica*, may grow to long, macroscopic visible bands, reaching lengths of more than
15 m and widths of 1–2 m, hanging down from the underside of the floes into the
water column (Melnikov & Bondarchuk 1987). Amphipods of the genera *Gammarus*,
Apherusa and *Onisimus* (Lønne & Gulliksen 1991) are permanently living at the
boundary between ice floes and the pelagic realm in densities of up to 60 individuals
per m^2 of ice (Carey 1985). These organisms, which are partially endemic to the

R. Gradinger

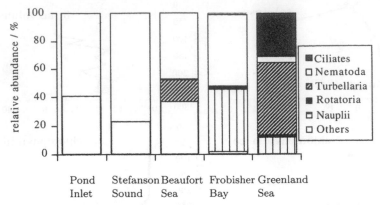

Figure 3. Relative composition of sea ice meiofauna in various parts of the Arctic Ocean. Data for Pond Inlet from Cross (1982); Stefanson Sound from Carey & Montagna (1982); Beaufort Sea from Kern & Carey (1983); Frobisher Bay from Grainger *et al.* (1985) and Greenland Sea from Gradinger *et al.* (1991).

Arctic Ocean, use the high algal biomass formed both directly at the underside and by the bottom community as a food source (Carey & Boudrias 1987). Besides the availability of food, they use the ice underside as a refuge to find shelter in the three-dimensional structure of, for example, pressure ridges.

Beside the autochthonous under-ice fauna, pelagic zooplankton, like the copepod species *Calanus glacialis* and *Pseudocalanus* spp. (specially *P. minutus*), temporarily ascend from deeper water layers to the underside of the ice floes to feed on ice algae (Runge *et al.* 1991). The under-ice fauna forms the link between the ice based primary production and the pelagic animals. These feed on ice algae and are important prey organisms for higher trophic levels like the polar cod (*Boreogadus saida*; Bradstreet & Cross 1982).

4. The pelagic realm

The biomass of pelagic organisms in the permanently ice-covered central regions of the Arctic Ocean is extremely low. The permanent ice cover reduces the incoming radiation, significantly suppressing algal growth to a degree already recognized by the early studies of Braarud (1935) and Steemann-Nielsen (1935). Concentrations of inorganic nutrients are relatively high throughout the year, and oxygen concentrations are in near equilibrium with the atmosphere, in agreement with the general idea of very low primary productivity in the central Arctic regions (Jones *et al.* 1990). The low algal biomass under the permanent pack ice is formed by small flagellates (Braarud 1935; Horner & Schrader 1982) in contrast to the diatom-dominated ice algal community. Investigations in the permanently ice-covered western part of the Greenland Sea (Gradinger & Baumann 1991) revealed an average algal biomass of 7 mg Chl m^{-2} in the upper 40 m of the water column under dense pack ice (figure 4), a value similar to the biomass observed inside the ice brine channel system. Thus, algal biomass has almost the same total value in sea ice and in the water column below, but the ambient concentrations (sea ice brine channels: greater than 400 mg Chl a m^{-3}; euphotic zone: less than 0.2 mg Chl a m^{-3}) are extremely different.

High phytoplankton concentrations, with integrated chlorophyll concentrations above 40 mg Chl m^{-2}, are restricted to marginal ice zones (MIZ) and polynyas. The

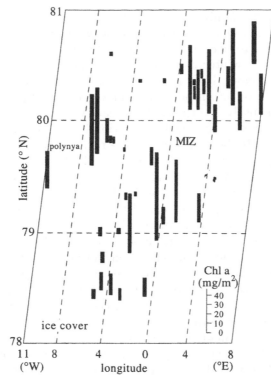

Figure 4. Distribution of algal biomass in the upper 40 m of the Greenland Sea (modified after Gradinger and Baumann (1991)).

significance of MIZ as regions of enhanced pelagic productivity was first shown for Arctic shelf areas (Roy & Loeng 1985; Alexander & Niebauer 1981), where melting of ice flocs leads (i) to an enhanced water column stratification and (ii) to increasing radiation. These conditions allow an even earlier onset of the phytoplankton growth in the MIZ than in the adjacent open water. Plankton blooms in MIZ are mainly formed by *Phaeocystis pouchetii* and pelagic diatom species (Gradinger & Baumann 1991). During the Arctic summer, nutrients become depleted in the upper layers of the water column (Spiess *et al.* 1988; Kattner & Becker 1991). Mesoscale processes like eddies and local wind-induced upwelling events (Buckley *et al.* 1979; Johannessen *et al.* 1983) lead to spatial patchiness in nutrient and algal concentrations and permit a prolongation of the algal growth period throughout the Arctic summer until the months September/October (Heimdal 1983).

Other areas of enhanced primary productivity in the Arctic Ocean are polynyas. The North East Water polynya, as one example, opens each year on the Greenland shelf, starting in late spring (May–June), and reaching its maximum extent of 44 000 km² in late summer (Wadhams 1981). Investigations in the polynya revealed similar biological characteristics to those described for marginal ice zones, since improved light availability and water column stratification enhance phytoplankton growth (Gradinger & Baumann 1991). The gradual increase in algal biomass is related to a decrease in nutrient concentrations until nitrate becomes depleted in the surface layer (Lara *et al.* 1994).

The life cycles of the Arctic zooplankton species are strongly adapted to the extreme seasonality and patchiness of food availability. During the short Arctic summer,

Arctic mesozooplankton, mainly consisting of copepods (*Calanus glacialis, Calanus hyperboreus*, and *Metridia longa*) feed and grow as young stages in the euphotic zone and accumulate energy-storage products, especially lipids, to survive the long starvation periods. They overwinter using a diapause-like strategy in deep waters, and again ascend to the euphotic layer in early or late spring (Smith & Schnack-Schiel 1990). The high algal biomass in polynyas and MIZ is used by the herbivorous zooplankton to sustain themselves in the Arctic Ocean. While the mesozooplankton may only have a minor impact on the algal production in the polynya and the marginal ice zone (Barthel 1986; Hirche *et al.* 1994) these regions are of special importance as areas of successful reproduction for the Arctic zooplankton (Hirche *et al.* 1991).

Due to the availability of food, marginal ice zones and polynyas are of major importance to the higher trophic levels of the Arctic marine ecosystem, as mesozooplankton species (*Calanus* spp.) are central to the pelagic food web (Bradstreet & Cross 1982). Various species of birds and marine mammals use marginal ice zones as migration routes due to the reliable availability of food (Ainley & DeMaster 1990). The breeding success of Arctic seabirds is dependent on the development of marginal ice zones at an accessible distance from the breeding grounds (Bradstreet 1988). Bird densities in marginal ice zones may be one to three orders of magnitude higher than in the adjacent ice-covered or open water area (Divoky 1979). Polynyas are for the same reasons attractors for both predators and their prey (Dunbar 1981). Large sea bird rookeries in the Canadian Arctic are located in the bird's flight range to a recurring polynya (Brown Nettleship 1981). The distribution of marine mammals is to a large extent determined by the position of polynyas as well (Stirling *et al.* 1981). Changes in the extent and distribution of polynyas, marginal ice zones and permanent ice cover will consequently directly influence recruitment success, migration behaviour and, in the long term, life cycle strategy of Arctic marine birds and mammals.

5. Effects of climate change on the Arctic marine system

The atmospheric CO_2 concentration has increased from a pre-industrial level of 280 ppm to a current level of approximately 360 ppm with an annual rate of about 1.5%. The concentration of methane, which is an even more effective greenhouse gas than CO_2, is increasing at a similar rate. These changes in the composition of the Earth's atmosphere have the potential to affect the global climate and increase the surface temperature. Predictions for the global climate in the next century using general circulation models indicate an enhanced warming of Arctic areas relative to lower latitudes, making the Arctic one of the most sensitive areas in the world.

Evidence of warming in high latitudes has already been gathered. Temperature in certain parts of northern Alaska has increased over the last decade, and will largely affect the conditions for terrestrial ecosystems (Oechel & Vourlitis 1994). The expected warming of the Arctic atmosphere will cause the permanent pack ice to shrink or even disappear in the next century.

A reduction in the extent of the permanent ice cover and a shift to a more seasonal ice regime will greatly affect the structure of the Arctic marine ecosystem (figure 5). Polynyas and marginal ice zones will occur in regions which until now have been characterized by a permanent ice cover and an extremely low biological productivity (Manak & Mysak 1989). Today, there is still uncertainty about the role of the biological CO_2 pump in relation to physical processes in general (Longhurst 1991) and in

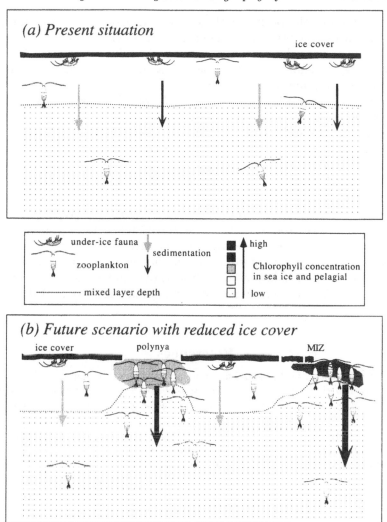

Figure 5. Structure of the marine ecosystem of the Arctic Ocean: (*a*) present state; (*b*) changes due to a reduction in the ice cover

polar oceans in particular, due to the scarcity of information (Legendre *et al.* 1992). Nevertheless, Anderson *et al.* (1990) have stated that the Arctic Ocean will be an active part of the biological pump transferring atmospheric CO_2 into the biogenic food web. An increase in the extent of polynyas and marginal ice zones further north will increase the biological productivity of the Arctic Ocean and the transfer of carbon from the atmosphere to the sea floor. Thus, the Arctic Ocean, despite its relatively small contribution to the world's ocean surface area, may play an important role in the global carbon cycle through enhancement of biological carbon fixation and subsequent sedimentation.

Besides the effects on total biological productivity, a reduction of the sea ice cover and changes in the location of polynyas and the marginal ice zones will have severe impact on several Arctic animals. Endemic ice-related species like the under-ice amphipod *Gammarus wilkitzkii*, or sea ice meiofauna species which are restricted in their distribution to the permanently ice-covered regions, will be diminished. Sea bird rookeries, located at present in the vicinity of polynyas and marginal ice zones, will either follow the receding ice extent, or the breeding success will decrease due to a higher energy consumption of the adults as a result of longer flight distances between feeding source and breeding area.

Endemic pelagic species like *Calanus glacialis* or *Calanus hyperboreus* will come into interspecific competition with sub-Arctic species like *Calanus finnmarchicus*, and the distribution boundaries of high-Arctic species may shift northward as sea-surface warming occurs. These changes in the composition of communities in the various habitats of the Arctic marine environment can be expected at timescales of years to decades and will largely depend on variations in the hydrographical regime, like, for example, the inflow of warm water from the North Atlantic.

The expected warming of the Arctic Ocean will change the structure of the marine communities into a more productive scenario. Harmful effects may be restricted to the flora and fauna living in close association to the Arctic multiyear ice floes. Greater danger to the Arctic marine environment on shorter timescales must be expected through pollution by oil, chlorinated hydrocarbons and radioactive waste, already introduced into the Arctic environment through human activity (Sakshaug & Skjodal 1989).

References

Ainley, D. G. & DeMaster, D. P. 1990 The upper trophic levels in polar marine ecosystems. In *Polar Oceanography* (ed. W. O. Smith Jr), pp. 599–630. San Diego: Academic Press.

Alexander, V. & Niebauer, H. J. 1981 Oceanography of the eastern Bering Sea ice-edge in spring. *Limnol. Oceanogr.* **26**, 1111–1125.

Anderson, L. G., Dyrssen, D. & Jones, E. P. 1990 An assessment of transport of atmospheric CO_2 into the Arctic Ocean. *J. Geophys. Res.* **95**, 1703–1711.

Assur, A. 1958 Composition of sea ice and its tensile strength. *Nat. Res. Council Publ.* **598**, 106–138.

Barthel, K.-G. 1986 Die Stellung dominanter Copepoden-Arten im Nahrungsgefüge typischer Wasserkörper der Grönland-See. *Ber. Inst. Meeresk.* **158**, 1–107.

Braarud, T. 1935 The 'Øst'-expedition to the Denmark Strait 1929. 2. The phytoplankton and its conditions of growth. *Hvalradets Skr.* **10**, 1–171.

Bradstreet, M. S. W. 1988 Importance of ice edges to high-Arctic seabirds. *Acta Congr. Int. Ornithol. 19th* **1**, 998–1000.

Bradstreet, M. S. M. & Cross, W. E. 1982 Trophic relationships at high Arctic ice edges. *Arctic* **35**, 1–12.

Brown, R. G. & Nettleship, D. N. 1981 The biological significance of polynyas to Arctic colonial seabirds. In *Polynyas in the Canadian Arctic* (ed. I. Stirling & H. Cleator), pp. 59–66. Ottawa: Canadian Wildlife Service.

Buckley, J. R., Gammelsröd, T., Johannessen, J. A., Johannessen, O. M. & Røed, L. P. 1979 Upwelling: oceanic structure at the edge of the Arctic ice pack in winter. *Science, Wash.* **203**, 165–167.

Carey, A. G. 1985 Marine ice fauna: Arctic. In *Sea ice biota* (ed. R. Horner), pp. 173–190. Boca Raton, FL: CRC Press.

Carey A. G. Jr & Boudrias, M. A. 1987 Feeding ecology of *Pseudalibrotus* (=*Onisimus*) *litoralis* Kroyer (Crustacea: Amphipoda) on the Beaufort Sea inner continental shelf. *Polar Biol.* **8**, 29–33.

Carey A. G. Jr & Montagna P. A. 1982 Arctic sea ice faunal assemblage: First approach to description and source of the underice meiofauna. *Mar. Ecol. Progr. Ser.* **8**, 1–8

Clarck, D. L. 1990 Stability of the Arctic Ocean ice-cover and Pleistocene warming events: Outlining the problem. In *Geological history of the polar oceans: Arctic versus Antarctic* (ed. U. Bleil & J. Thiede), pp. 273–287. Dordrecht: Kluwer.

Clasby, R., Horner, R. & Alexander, V. 1973 An *in situ* method for measuring primary production of Arctic sea ice algae. *J. Fish. Res. Board Can.* **30**, 635–638.

Cross, W. E. 1982 Under-ice biota at the Pond Inlet ice edge and in adjacent fast ice areas during spring. *Arctic* **35**, 13–27.

Divoky, G. J. 1979 Sea ice as a factor in seabird distribution and ecology of the Beaufort, Chukchi, and Bering seas. In *Conservation of marine birds of northern North America* (ed. J. C. Bartonek & D. N. Nettleship), pp. 9–18. Washington: US Fisheries Wildlife Service.

Dunbar, M. J. 1981 Physical causes and biological significance of polynyas and other open water in sea ice. In *Polynyas in the Canadian Arctic* (ed. I. Stirling, & H. Cleator) pp. 29–44 Ottawa: Canadian Wildlife Service.

Frankenstein, G. & Garner, R. 1967 Equations for determining the brine volume of sea ice from $-0.5\,°C$ to $-22.9\,°C$. *J. Glaciol.* **6**, 943–944.

Gradinger, R. R. & Baumann, M. E. M. 1991 Distribution of phytoplankton communities in relation to large-scale hydrographical regime in the Fram Strait. *Mar. Biol.* **111**, 311–321.

Gradinger, R., Spindler, M. & Henschel, D. 1991 Development of Arctic sea-ice organisms under graded snow cover. *Polar Res.* **10**, 295–308.

Grainger, E. H. 1979 Primary production in Frobisher Bay, Arctic Canada. In *Marine production mechanisms* (ed. M. J. Dunbar), pp. 9–30. Cambridge: Cambridge University Press.

Grainger, E. H., Mohammed, A. A. & Lovrity, J. E. 1985 The sea ice fauna of Frobisher Bay, Arctic Canada. *Arctic* **38**, 23–30.

Heimdal, B. R. 1983 Phytoplankton and nutrients in the waters north-west of Spitsbergen in the autumn of 1979. *J. Plankton Res.* **5**, 901–918.

Hirche, H. -J., Baumann, M. E. M., Kattner, G., & Gradinger, R. 1991 Plankton distribution and the impact of copepod grazing on primary production in Fram Strait, Greenland Sea. *J. Mar. Syst.* **2**, 477 494.

Hirche, H. -J., Hagen, W., Mumm, N. & Richter, C. 1994 The Northeast Water Polynya, Greenland Sea. III. Meso- and macrozooplankton distribution and production of dominant herbivorous copepods during spring. *Polar Biol.* **14**, 491–503

Horner, R. 1985 *Sea Ice Biota*. Boca Raton, FL: CRC press.

Horner, R. & Schrader, G. C. 1982 Relative contribution of ice algae, phytoplankton, and benthic microalgae to primary production in nearshore regions of the Beaufort Sea. *Arctic* **35**, 485–503.

Johannessen, O. M., Johannessen, J. A., Morison, J., Farrelly, B. A. & Svendsen, E. A. S. 1983 Oceanographic conditions in the MIZ north of Svalbard in early fall 1979 with emphasis on mesoscale processes. *J. Geophys. Res.* **88**, 2755–2769.

Jones, E. P., Nelson, D. M. & Treguer, P. 1990 Chemical Oceanography. In *Polar oceanography* (ed. W. O. Smith Jr), pp. 407–476. San Diego: Academic Press.

Kattner, G. & Becker, H. 1991 Nutrients and organic nitrogenous compounds in the MIZ of the Fram Strait. *J. Mar. Syst.* **2**, 385–394.

Kern, J. C. & Carey, A. G. Jr 1983 The faunal assemblage inhabiting seasonal sea ice in the nearshore Arctic Ocean with emphasis on copepods. *Mar. Ecol. Progr. Ser.* **10**, 159–167.

Kurbjeweit, F., Gradinger, R. & Weissenberger, J. 1993 The life cycle of *Stephos longipes* – an example for cryopelagic coupling in the Weddell Sea (Antarctica). *Mar. Ecol. Progr. Ser.* **98**, 255–262.

Lara, R. J., Kattner, G. & Tillmann, U. 1994 The North East Water polynya (Greenland Sea) II. Mechanisms of nutrient supply and influence on phytoplankton distribution. *Polar Biol.* **14**, 483–490.

Legendre, L., Ackley, S. F., Dieckmann, G. S., Gulliksen, B., Horner, R., Hoshiai, T., Melnikov, I. A., Reeburgh, W. S., Spindler, M. & Sullivan, C. W. 1992 Ecology of sea ice biota. 2. Global significance. *Polar Biol.* **12**, 429–444.

Longhurst, A. R. 1991 A reply to Broecker's charges. *Global Biogeochem. Cycles* **5**, 315–316.

Lønne, O. J. & Gulliksen, B. 1991 On the distribution of sympagic macro-fauna in the seasonally ice covered Barents Sea. *Polar Biol.* **11**, 457–469.

Manak, D. K. & Mysak, L. 1989 On the relationship between Arctic sea ice anomalies and fluctuations in northern Canadian air temperature and river discharge. *Atmos. Ocean* **27**, 682–691.

Maykut, G. A. 1985 The ice environment. In *Sea ice biota* (ed. R. Horner), pp. 21–82. Boca Raton, FL: CRC Press.

Melnikov, I. A. & Bondarchuk, L. L. 1987 Ecology of mass accumulations of colonial diatom algae under drifting Arctic ice. *Oceanology* **27**, 233–236.

Mitchell, J. F. B., Manabe, S., Tokioka, T. & Meleshko, V. 1990 Equilibrium climate change In *Climate change: the IPCC scientific assessment* (ed. J. T. Houghton, G. J. Jenkins & J. J. Ephraums), pp. 131–172. Cambridge University Press.

Oechel, W. C. & Vourlitis, G. L. 1994 The effects of climate change on land-atmosphere feedbacks in Arctic tundra regions. *TREE* **9**, 324–329.

Poulin, M. 1990 Ice diatoms: the Arctic. In *Polar marine diatoms* (ed. L. K. Medlin & J. Priddle), pp. 15–18 Cambridge: British Antarctic Survey.

Rey, F. & Loeng, H. 1985 The influence of ice and hydrographic conditions on the development of phytoplankton in the Barents Sea. In *Marine biology of polar regions and effects of stress on marine organisms* (ed. J. S. Gray & M. E. Christiansen), pp. 49–63. Chichester: Wiley.

Runge, J. A., Therriault, J., Legendre, L., Ingram, R. G. & Demers, S. 1991 Coupling between ice microalgal productivity and the pelagic, metazoan food web in southeastern Hudson Bay: a synthesis of results. *Polar Res.* **10**, 325–338.

Sakshaug, E. & Skjodal, H. R. 1989 Life at the ice edge. *Ambio* **18**, 60–67.

Smith, R. E. H., Clement, P. & Head, E. 1989 Biosynthesis and photosynthate allocation patterns of Arctic ice algae. *Limnol. Oceanogr.* **34**, 591–605.

Smith, S. L. & Schnack-Schiel, S. B. 1990 Polar zooplankton. In *Polar oceanography* (ed. W. O. Smith Jr), pp. 527–598. San Diego: Academic Press.

Spies, A., Brockmann, U. H. & Kattner, G. 1988 Nutrient regimes in the MIZ of the Greenland Sea in summer. *Mar. Ecol. Progr. Ser.* **47**, 195–204

Spindler, M. 1990 A comparison of Arctic and Antarctic sea ice and the effects of different properties on sea ice biota. In *Geological history of the polar oceans: Arctic versus Antarctic* (ed. U. Bleil & J. Thiede), pp. 173–186. Dordrecht: Kluwer.

Steeman-Nielsen, E. 1935 The production of phytoplankton of the Faroe Isles, Iceland, East Greenland and in the waters around. *Komm. Dan. Fisk. Havundersög. Medd. Ser. Plankton* **3**, 1–93.

Stirling, I., Cleator, H. & Smith TG 1981 Marine mammals. In *Polynyas in the Canadian Arctic* (ed. I. Stirling & H. Cleator), pp. 59–66 Ottawa: Canadian Wildlife Service.

Wadhams P. 1981 The ice cover in the Greenland and Norwegian Sea. *Rev. Geophys. Space Physics* **19**, 345–393.

Weissenberger, J., Dieckmann, G., Gradinger, R. & Spindler, M. 1992 Sea ice: A cast technique to examine and analyze brine pockets and channel structure. *Limnol. Oceanogr.* **37**, 179–183.

The thermohaline circulation of the Arctic Ocean and the Greenland Sea

By Bert Rudels

Institut für Meereskunde der Universität Hamburg,
Troplowitzstraße 7, D-22529 Hamburg, Germany

The thermohaline circulation of the Arctic Ocean and the Greenland Sea is conditioned by the harsh, high latitude climate and by bathymetry. Warm Atlantic water loses its heat and also becomes less saline by added river run-off. In the Arctic Ocean, this leads to rapid cooling of the surface water and to ice formation. Brine, released by freezing, increases the density of the surface layer, but the ice cover also insulates the ocean and reduces heat loss. This limits density increase, and in the central Arctic Ocean a low salinity surface layer and a permanent ice cover are maintained. Only over the shallow shelves, where the entire water column is cooled to freezing, can dense water form and accumulate to eventually sink down the continental slope into the deep ocean. The part of the Atlantic water which enters the Arctic Ocean is thus separated into a low density surface layer and a denser, deep circulation. These two loops exit through Fram Strait. The waters are partly rehomogenized in the Greenland Sea. The main current is confined to the Greenland continental slope, but polar surface water and ice are injected into the central gyre and create a low density lid, allowing for ice formation in winter. This leads to a density increase sufficient to trigger convection, upwelling and subsequent ice melt. The convection maintains the weak stratification of the gyre and also reinforces the deep circulation loop. As the transformed waters return to the North Atlantic the low-salinity, upper water of the East Greenland Current enters the Labrador Sea and influences the formation of Labrador Sea deep water. The dense loop passes through Denmark Strait and the Faroe–Shetland Channel and sinks to contribute to the North Atlantic deep water. Changes in the forcing conditions might alter the relative strength of the two loops. This could affect the oceanic thermohaline circulation on a global scale

1. Introduction

The presence of the Arctic Mediterranean Sea north of the Greenland–Scotland Ridge allows warm water from the Atlantic to reach the shores of northern Europe and the continents bordering on the Arctic Ocean. The warm Atlantic water in the Norwegian Sea strongly influences the climate of northwestern Europe. The effects are largely removed east of the Barents Sea and almost absent in the Arctic Ocean. Here, in contrast, a severe climate dominates and determines the oceanic conditions. To a certain extent, the Arctic Ocean–Greenland Sea is a cul-de-sac (figure 1), where water mass transformations, some affecting the global oceanic thermohaline circulation, take place.

The heat loss and the large fresh water input, mainly as river run-off, lead to a

Figure 1. The surface circulation in the Arctic Ocean and the Greenland Sea. (1) Lomonosov Ridge, (2) Canadian Basin, (3) Eurasian Basin, (4) Amundsen Basin, (5) Gakkel Ridge, (6) St Anna Trough, (7) Norwegian Atlantic Current, (8) West Spitsbergen Current, (9) Beaufort Gyre, (10) Transpolar Drift, (11) East Greenland Current.

cooling of the Atlantic water and to an increase in the stability of the water column, which permits a reduction of the upper layer temperatures to the freezing point. An ice cover forms, seasonal over the shelves, but perennial in the central Arctic Ocean. Freezing leads to ejection of brine which increases the salinity and density of the underlying water. The highest salinities are reached on the shallow shelf seas, especially in areas of frequent open water, where the shelf water is transformed into low salinity surface water in the run-off dominated summer and into saline, dense waters in winter. These sink off the shelves down the continental slope as entraining density flows supplying the deeper layers.

The Arctic waters are thus derived from interactions between Atlantic water and river run-off. This implies a splitting of the inflow into a low salinity surface loop and a deep water loop consisting of cooled Atlantic and denser waters.

The transformed waters and the sea ice exit the Arctic Ocean through Fram Strait and enter the Greenland Sea, the second area of water formation and deep convection. Open ocean convection cools the deep waters and diminishes their temperature–salinity (Θ–S) range. The upper loop, consisting of polar surface water and ice, is partly instrumental for, by creating a low salinity surface layer in the central Greenland Sea, but largely unaffected by the convection.

The two loops recross the Greenland–Scotland Ridge. The denser loop supplies, as a deep boundary current, part of the North Atlantic deep water, while the upper loop flows around Greenland into the Labrador Sea and Baffin Bay as the West Greenland Current. This inflow of low salinity water influences the production and the characteristics of the Labrador Sea deep water, a second source contributing to the North Atlantic deep water (McCartney & Talley 1984). The processes in the Arctic Ocean and the Greenland Sea thus affect and partly drive the global oceanic thermohaline circulation, the main ventilation of the deep waters of the oceans.

This simple picture has obvious limitations, the most critical being the neglect of the Pacific inflow through Bering Strait. However, a corresponding outflow, in volume as well as salt, occurs through the Canadian Arctic Archipelago and for a zero order approximation this through flow can be considered decoupled from the Atlantic circulation north of the Greenland–Scotland Ridge. The low salinity Pacific inflow is then directly involved in the formation of Labrador Sea deep water and becomes drawn into the deep thermohaline circulation. The Pacific inflow will not be considered any further.

In the following sections the routes of the Atlantic water, the boundary convection, the water mass transformation and the circulation in the Arctic Ocean are described. The convection processes in the Greenland Sea and the importance of the interactions with the Arctic and Atlantic waters for the Greenland Sea water column are discussed. The implications of the water transformations for the global thermohaline circulation are then addressed.

This is not a review but represents a personal view, developed over the last five years, of the processes active in the Arctic Mediterranean. For a summary of Arctic oceanography Coachman & Aagaard (1974), Carmack (1986, 1990), Jones *et al.* (1990) and Muench (1990) should be consulted.

2. The Atlantic inflow

Warm Atlantic water crosses the Greenland–Scotland Ridge. The inflow is estimated to be 5–8 Sv (10^6 m^3 s^{-1}) (Worthington 1970; McCartney & Talley 1984). It flows as the Norwegian Atlantic Current until it reaches the latitudes of the Barents Sea. There it splits. One part enters together with the Norwegian Coastal Current the Barents Sea, while the outer part continues as the West Spitsbergen Current toward Fram Strait. Again the current splits. A small fraction (1 Sv, Bourke *et al.* 1988) enters the Arctic Ocean, while the main part recirculates in several branches towards the west (Quadfasel & Meincke 1987). The Atlantic water is cooled on its way toward the north and the winter convection in the Norwegian Sea homogenizes the water column down to 600–800 m.

The Atlantic water entering the Arctic Ocean flows as a boundary current along the continental slope toward the east. It interacts strongly with sea ice north of Svalbard and a less saline surface water is formed, which becomes homogenized by freezing and convection in winter into a deep mixed layer. This layer appears to follow the Atlantic water as a protective lid, shielding it from the surface processes (Rudels *et al.* 1995a). Later transformations of the Atlantic Layer occur through interactions with dense water leaving the shelves.

The Barents Sea inflow is subject to stronger exchanges with the atmosphere and its density range is expanded. In its upper part it becomes colder, less saline and less dense, while in the deeper part the water becomes colder and denser, by cooling and

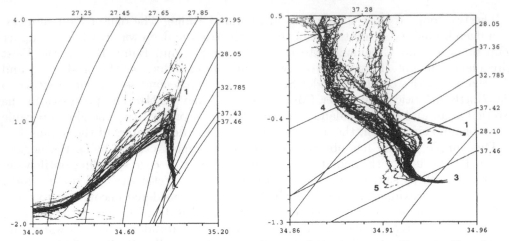

Figure 2. Θ–S diagram showing stations taken from IB Oden during the Arctic-91 Expedition. (a) water masses below the halocline. (1) Atlantic layer and inversions. (b) Blow-up of the deep waters. (1) Canadian Basin stations, (2) Canadian Basin deep water salinity maximum in the Eurasian Basin, (3) salinity maximum of Eurasian Basin deep water, (4) stations in the Amundsen Basin, (5) trace of Norwegian Sea deep water north of the Yermak Plateau. Note also the inversions in the temperature range $-0.5 < Q < 0$ of the upper polar deep water.

by incorporating brine enriched water formed over the shallow areas of the Barents Sea, possibly west of Novaya Zemlya. It passes between Frans Josef Land and Novaya Zemlya and sinks down the St Anna Trough into the Arctic Ocean, where it forms a deep (greater than 1000 m) wedge at the continental slope. The two inflow branches meet north of the Kara Sea and continue in the boundary current eastward (Rudels *et al.* 1994).

3. Circulation and water transformation in the Arctic Ocean

The surface layer is, further to the east, supplied by injections of low salinity water from the shelves. A polar mixed layer is established above the water homogenized north of Svalbard, which now forms a halocline isolating the Atlantic Layer from the polar mixed layer. The halocline also becomes decoupled from surface processes, and can only be replenished by injections of dense water from the shelves in winter (Aagaard *et al.* 1981).

The circulation of the polar mixed layer is anticyclonic and dominated by the wind driven Beaufort Gyre. The transpolar drift moves ice and low salinity surface water out of the Beaufort Gyre and across the Lomonosov Ridge close to the North Pole. The Siberian branch of the drift flows from the Laptev Sea northward but then veers toward Fram Strait. The motions in the upper layers in the southern part of the Eurasian Basin are less certain, but there are indications of an eastward flow in the boundary current above the Atlantic Laye.

The transformation of the Atlantic water in the Arctic Ocean can be inferred by examining Θ–S curves of hydrographic stations. Figure 2 shows stations occupied by IB Oden in the Arctic-91 expedition (Anderson *et al.* 1994). These give, not an exhaustive, but a fair representation of the Arctic Ocean waters below the halocline.

The previously smooth Θ–S curve of the Atlantic water exhibits inversions in temperature and salinity, and the temperature of the Atlantic Layer is reduced by incorporating colder, less saline water. In the Canadian Basin only dense shelf water can penetrate deeper than 200 m and interact with the water from the Eurasian Basin crossing the Lomonosov Ridge. The Atlantic Layer (200–700 m) is colder and the intermediate depth layer (700–1700 m) is warmer than in the Eurasian Basin (figure 2). Assuming that the water entering the Canadian Basin has characteristics similar to those found in the Amundsen Basin close to the Lomonosov Ridge, it is obvious that the boundary convection from the shelves partly enters and cools the Atlantic Layer, partly redistributes, by entraining Atlantic water, heat downward to deeper layers.

On the other side of the Lomonosov Ridge, in the Amundsen Basin, colder, not warmer, water has been added to the intermediate depth layers (figure 2). This can only happen if the entering water does not sink through the warm Atlantic Layer of the boundary current. It therefore implies an inflow strong enough to push the Atlantic water away from the slope (Rudels *et al.* 1994).

The waters of the Amundsen Basin and over the Gakkel Ridge also display inversions, strong in the warm Atlantic core and weaker but very regular at the intermediate depth below (figure 2). The upper inversions could be due to intrusions of dense water from the shelves (Quadfasel *et al.* 1993), as well as by the Barents Sea inflow, but the regular, deeper lying inversions indicate interactions across a narrow front over an extended depth interval (Rudels *et al.* 1994). The inversions are found far from the Eurasian continental slope and must have been advected with the mean flow. They can then be used as markers for the circulation (Quadfasel *et al.* 1993).

The Atlantic inflow over the Barents Sea provides a strong, cold injection into the Arctic Ocean water column. Recent current measurements between Frans Josef Land and Novaya Zemlya (Loeng *et al.* 1993) have shown that an inflow of 2 Sv, almost twice the inflow through Fram Strait, enters the Kara Sea. The existence of a colder, low salinity wedge close to the continental shelf north of the Laptev Sea has also been observed (Schauer *et al.* 1995).

The two inflows, from Fram Strait and from the Barents Sea, meet north of the Kara Sea. They merge across a narrow front and create inversions in temperature and salinity over an extended depth range, as they continue eastward. The boundary current then branches north of the Laptev Sea. The larger fraction returns toward Fram Strait with the outer, warmer branch dominating over the Gakkel Ridge and the colder Barents Sea branch being more prominent closer to the Lomonosov Ridge. A smaller part of the boundary current crosses the Lomonosov Ridge and enters the Canadian Basin (Rudels *et al.* 1994).

In the deepest layers the Θ–S curves of the two basins change their relative slopes. The deep and bottom waters in the Eurasian Basin show a salinity increase and an almost constant temperature, while in the Canadian Basin the salinity of the deepest layers remains constant, with the temperature decreasing (figure 2). Boundary convection from the shelves leads to high salinities and to constant temperatures at the deepest levels, since the initial temperature of the shelf waters is the same and they all pass through the boundary current and entrain waters of similar properties. This appears to occur in the deep Eurasian Basin. By contrast, the decreasing temperature in the deep Canadian Basin suggests that, in addition to the boundary current along the Siberian continental slope, Eurasian Basin waters pass through rifts in the central part the Lomonosov Ridge. This spillover would sink toward the bottom entraining ambient water just as the slope convection, and it would add colder water to the deep Canadian Basin.

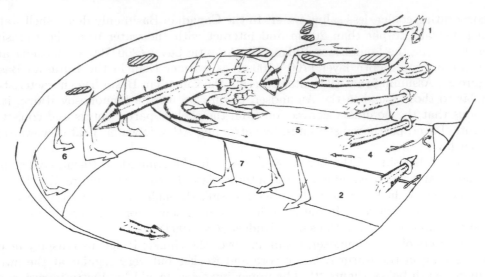

Figure 3. Hypothetical picture of the circulation in the Arctic Ocean. (1) Fram Strait, (2) Lomonosov Ridge, (3) Atlantic and intermediate depth circulation, (4) circulation of Canadian Basin deep water, (5) circulation of Eurasian Basin deep water, (6) slope convection, (7) convection down the Lomonosov Ridge.

To determine the Θ–S characteristics of the water added to the water column below 200 m in the Canadian Basin a simple mass balance model has been applied (Rudels *et al.* 1994; Jones *et al.* 1995). It is assumed that an ensemble of thin, transient plumes leaves the shelves. The plumes entrain ambient water and enter, when they have reached a depth corresponding to their density, into the water column. They merge in the upper part (above 1700 m) with the water of the boundary current from the Eurasian Basin to form the Canadian Basin water column. Below 1700 m the boundary current cannot cross the Lomonosov Ridge. The spillover across the central part of the ridge provides, for the deeper layers, the water mass which can balance the now warm and saline shelf-slope contribution to form the lower part of the Canadian Basin water column. To reproduce the Canadian Basin characteristics a high entrainment rate has to be assumed and the convecting plumes, if they reach the deepest part of the basin, have increased their volume by a factor of 20.

The Canadian Basin waters recross the Lomonosov Ridge north of Greenland and the Canadian Basin deep water can be identified in the Eurasian Basin as a salinity maximum at about 1800 m (figure 2). It is strongest close to the Morris Jesup Plateau, but it is also seen in the Amundsen Basin away from the Lomonosov Ridge. This implies a splitting of the Canadian Basin deep outflow north of Fram Strait. One part flows below and against the Atlantic and intermediate layers into the Amundsen Basin, while the other part exits through Fram Strait along the Greenland continental slope. The circulation of the Atlantic, intermediate depth and deep layers is sketched in figure 3.

4. Deep water renewal in the Greenland Sea

The Arctic Ocean waters exit through Fram Strait, where they meet recirculating Atlantic water of the West Spitsbergen Current. The outflow through Fram Strait is about 3 Sv, 1 Sv polar surface water, 1 Sv Atlantic water and 1 Sv of intermediate and

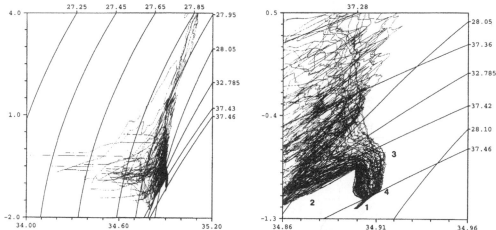

Figure 4. Θ–S diagram showing stations obtained by RV Valdivia in the Greenland Sea in May 1993. (a) The entire water column. (b) Blow-up of the deep waters. (1) Greenland Sea bottom water, (2) Arctic intermediate water, (3) deep temperature maximum, (4) deep salinity maximum.

deep waters, while the strength of the recirculation of Atlantic water in Fram Strait is 1–2 Sv (Bourke *et al.* 1988; Rudels 1987) The Θ–S characteristics of the water masses of the Greenland Sea are shown in figure 4 which displays stations obtained during the Valdivia 136 cruise in 1993 (Rudels *et al.* 1993). The waters are clearly distinct from those of the Arctic Ocean. They are colder and less saline. The intermediate upper polar deep water from the Arctic Ocean, identified by salinities and temperatures increasing and decreasing respectively with depth (figure 2), contrasts strongly with the Arctic Intermediate water of the Greenland Sea in the same density range but with salinities and temperatures both increasing with depth. The deep waters of the Greenland Sea, in fact, occupy a large part of the empty Θ–S space below the Arctic Ocean Θ–S curves.

However, certain features of the Arctic Ocean deep waters are recognized also in the Greenland Sea. The deep temperature maximum can be associated with the salinity maximum of the Canadian Basin deep water in the Eurasian Basin and the deep salinity maximum in the Greenland Sea indicates that some of the Eurasian Basin deep water passes through the 2600 m deep Fram Strait and enters the Greenland Sea (figures 4 and 2).

The cooling of the waters from the north occurs through open ocean convection and the associated transport of cold surface water into the deep. Convection, especially when ice formation is involved, occurs on small (0.1–1 km) scales and during short periods. Its observation and description require sophisticated measuring techniques and the inclusion of small scale processes and nonlinear effects into high resolution non-hydrostatic models. (Backhaus 1995; Garwood 1991; Garwood *et al.* 1995; Jones & Marshall 1993; Latarius & Quadfasel, personal communication; Rudels 1990; Rudels & Quadfasel 1991; Schott *et al.* 1993). A discussion of this small scale work is not attempted here. Instead we focus on integral effects and on how the presence of sea ice influences the evolution of the mixed layer and the conditions for deep convection (Walin 1993). A simple one-dimensional energy balance ice-mixed layer model is examined. The energy input necessary for entrainment is supplied by

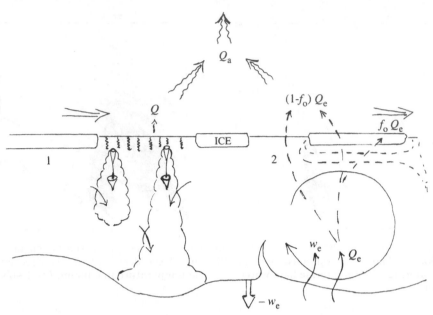

Figure 5. Description of the mixed layer processes in the presence of sea ice during cooling. (1) Freezing and convection, (2) entrainment and melting. These processes occur simultaneously. Q_a is the heat loss to the atmosphere and Q the latent heat released by freezing. w_e is the entrainment velocity and $-w_e$ the deepening of the mixed layer. f_o is the fraction of entrained heat going to ice melt.

the wind and a constant cooling rate is assumed. The presentation below follows Rudels et $al.$ (1995b).

The densities required for deep convection are normally not reached by cooling alone, and an initial ice formation takes place. The release of brine lowers the stability at the base of the mixed layer, and water from below is more easily entrained. The mixed layer deepens and the enhanced vertical heat flux reduces the ice formation. The density of the mixed layer thus increases, partly because of the entrainment and the cooling of water from below and partly by the release of brine by freezing (figure 5).

Some of the entrained heat goes to ice melt, and thus provides a positive buoyancy flux at the sea surface directly coupled to the entrainment. It can be shown that there exists a fraction,

$$f_o = 2\alpha L/(c(\beta S_2 - \alpha \Delta T)) \approx 0.23,$$

for which the ice melt rate is a minimum (Rudels et $al.$ 1995). Here c and L are the heat capacity and the latent heat of melting of sea water; α and β are the coefficients of heat expansion and salt contraction; S_2 is the salinity of the lower layer and ΔT the temperature difference between the mixed layer and the underlying water. If the entrained heat is assumed to be distributed in this way, the entrainment velocity remains finite, while the stability at the mixed layer base goes to zero. The mixed layer eventually becomes denser than the underlying water and convects into the deep.

The interface rises toward the surface as the mixed layer is removed. The entrainment increases with decreasing mixed layer depth, and when a balance between heat loss to ice melt and to the atmosphere and entrained heat is reached the convection ceases and a new mixed layer is established. It has temperatures above freezing and at its base the destabilizing density step due to temperature is half the stabilizing density step due to salinity.

A period of net melting follows as the mixed layer again deepens and its temperature decreases. The ratio of the saline to thermal density steps, however, remains constant. If the freezing point is reached before the ice cover is removed, this coupling between the density steps is broken. A new cycle of net freezing and convection then occurs. The number of convection events depends upon the fresh water content of the mixed layer and the temperature of the underlying water. If a low salinity mixed layer is present and the underlying layer is cool, the haline convection persists longer and several convection events occur. With a higher salinity in the mixed layer and warm Atlantic water below the ice will be removed after one event. When all ice has melted, the mixed layer temperature is above freezing and the stability at its base so weak that no more ice can form and only thermal convection takes place during the rest of the winter. Ice-free conditions are by far the most frequently occuring final situations.

Cooling does not give rise to as large density anomalies at the sea surface as freezing, and the suppression of entrainment due to ice melt is removed. A marginally stable mixed layer, gradually deepening by a combination of thermal convection and entrainment then replaces the sudden emptying of the mixed layer occurring during haline convection (Backhaus 1995; Houssais & Hibler 1993).

The warm, saline situation has been the one most often encountered in the Greenland Sea in recent years and an extreme case occurred in 1994, when no ice was formed in the central Greenland Sea and only thermal convection down to 600 m took place (Latarius & Quadfasel, personal communication). The convective regime in the Greenland Sea then resembles the winter deepening in the Norwegian Sea and in the high latitude branches of the subpolar and subtropical gyres (McCartney 1982).

This situation is due to larger injections of Atlantic water and smaller injections of polar water from the East Greenland Current into the central Greenland Sea, which reduce the lifetime of the ice cover and initiate the thermal convection at an early stage. The deepest layers are then not ventilated because high enough surface densities are not reached (figure 6). The density of the central gyre is reduced and the doming of the isopycnals cannot be maintained. The water column relaxes and slumps towards the rim of the basin. This allows for a penetration of the Arctic Ocean deep waters toward the centre of the gyre and distinct Arctic features such as the deep temperature and salinity maxima become more prominent. The penetration occurs isopycnally and the density of the maxima does not change, but because of the relaxation of the dome the maxima are displaced downward (Rudels *et al.* 1993; Meincke & Rudels 1995). The doming of the Greenland Sea gyre then appears partly to be a thermohaline, not just a wind generated feature.

The circulation of the deepest layers is internal to the Arctic Mediterranean (Aagaard *et al.* 1985; Rudels 1986; Rudels & Quadfasel 1991). The merging of Greenland Sea deep water with the Canadian and Eurasian Basin deep water on the Greenland continental slope forms the Norwegian Sea deep water, which is injected along the Jan Mayen Fracture Zone into the Norwegian Sea (Aagaard *et al.* 1985). Water with Norwegian Sea deep water characteristics is also formed by isopycnal mixing in Fram Strait. Norwegian Sea deep water has been assumed to be the principal deep water component which enters the Arctic Ocean from the south. However, only weak in-

B. Rudels

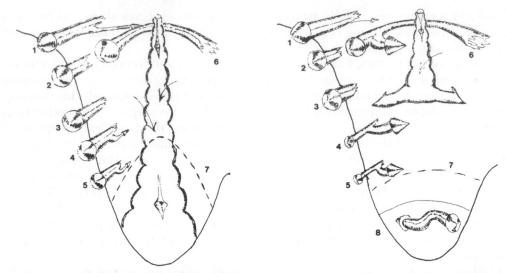

Figure 6. Hypothetical picture of circulation in the Greenland Sea. (*a*) Convection to the bottom and formation of bottom water. (*b*) Convection down to intermediate depth and formation of Arctic intermediate water. (1) Polar surface water, (2) water from the Atlantic Layer, (3) upper polar deep water, (4) Canadian Basin deep water, (5) Eurasian Basin deep water, (6) recirculating Atlantic water, (7) isopycnal surface, (8) spin down of Greenland Sea bottom water. While the different waters still can be identified, their original $\Theta-S$ characteristics have been greatly removed, predominantly by isopycnal mixing, during their transits toward the Greenland Sea.

dications of Norwegian Sea deep water can be seen north of the Yermak Plateau (figure 2). Even if deep convection has not occurred recently in the Greenland Sea, the formation of Arctic Intermediate water is active. This convection does not reach deep enough to incorporate the Canadian Basin deep water, but it is dense enough to reinforce the intermediate depth layers of the East Greenland Current. It will merge with the Arctic Ocean outflow and also with the recirculating Atlantic water and reduce their salinities and temperatures as they move toward Denmark Strait. Further additions to the East Greenland Current occur in the Icelandic Sea, where a warmer mode of Arctic Intermediate water is formed (Swift & Aagaard 1981; Swift *et al.* 1980).

5. Variability of the Arctic Ocean–Greenland Sea circulation

The deep return loop crosses the Greenland-Scotland Ridge through Denmark Strait and the Faroe–Shetland Channel (its densest part) and contributes to the formation of North Atlantic deep water and the driving of the global thermohaline circulation. Several processes add to the overflow water: the boundary convection in the Arctic Ocean; the cooling of the inflow in the Barents Sea; the open ocean convection in the Greenland and Icelandic Seas; and the cooling, by isopycnal mixing, of the recirculating Atlantic water (Strass *et al.* 1993).

The existence of several sources for the overflow into the North Atlantic implies that if one of the sources is reduced, the others may fill the deficit. Because of its large volume and buffer capacity and its inaccessibility to observations variations in the Arctic Ocean sources are difficult to detect. The situation is different in the smaller

Greenland Sea, and recent research has indicated large changes in the convection and deep water ventilation (GSP group 1990).

The convective regime has varied between a low salinity surface layer with a shallow, haline convection down to 200–300 m in 1982 (Clarke *et al.* 1990); and a high salinity, deep mixed layer, no ice and thermal convection down to 600 m in 1994 (Latarius & Quadfasel, personal communication). In the intervening years convections down to 1300 m in 1988 (Rudels *et al.* 1989) and 2000 m in 1989 (GSP group 1990) were observed. The event in 1988 could have been caused by haline convection, while the final deepening in 1989 was thermal (Fisher, personal communication).

No convection to the bottom has occurred recently and the temperature and salinity of the Greenland Sea bottom water has increased in the last 10 years (Meincke *et al.* 1992). The change is even more striking, if we compare with the bottom temperatures observed in the first part of the century (Kiilerich 1945), below $-1.4\,°C$ as compared to just below $-1.2\,°C$.

Two conditions are required (in my opinion) to form Greenland Sea bottom water: a sufficiently weak stability at the base of the mixed layer for the haline convection to break out of the mixed layer; and underlying water cool enough not to change the haline convection into a thermal one. The Greenland Sea has gradually been filled with Arctic Ocean deep waters and assumed a more Arctic character (Rudels *et al.* 1993; Meincke & Rudels 1995). This also implies that the deep water contribution from the south to the Arctic Ocean is reduced and the internal deep water circulation in the Arctic Mediterranean is becoming weaker.

The reduction of the Greenland Sea deep water formation has not affected the formation of Arctic Intermediate water and the supply to the overflow has been maintained. Only in the case of a stable, low salinity upper layer and a shallow convection as in 1982 could the Greenland Sea contribution be removed. This appears to have happened on a grand scale in connection with the 'great salinity anomaly' (Dickson *et al.* 1988). Then not only the production of Arctic Intermediate water in the Greenland and Icelandic Seas was shut off. The low salinity water also entered the Labrador Sea and closed down the formation of Labrador Sea deep water. Both the overflow and the Labrador Sea contribution to the North Atlantic deep water were then reduced.

The origin of the salinity anomaly is not established. Was it a local effect in the Greenland Sea, or did it also signal changes occurring in the Arctic Ocean? The critical part in the Greenland Sea is the East Greenland Current. Why does it remain stable and restrict the outflow of low salinity polar surface water to a buoyant boundary current above the continental slope? If a perturbation of the East Greenland Current occurred, an injection of low salinity water into the central Greenland Sea would affect all downstream sources of the North Atlantic deep water.

However, if the reduced salinity were due to more fresh water in the Arctic Ocean water column, it could indicate more drastic changes. Naively, a higher fresh water content in the water column suggests lower salinities on the Arctic Ocean shelves. This leads to a reduction of dense water formation during winter. A smaller amount of Arctic Ocean deep waters would then be formed and all sources of the overflow water would be reduced. The global thermohaline circulation is then likely to be affected.

Could a partial melting of the Arctic Ocean ice cover be sufficient to bring about these changes or is an increased northward atmospheric fresh water flux required? Such questions demand a more comprehensive view of the climate system than can be obtained by examining only the thermohaline circulation in the Arctic Ocean and the Greenland Sea.

My understanding of the oceanography of the Arctic Ocean and the Greenland Sea has developed during constant discussion with: L. Anderson, J. Backhaus. H. Friedrich, M.-N. Houssais, E. P. Jones, J. Meincke and D. Quadfasel. Financial support for this work was granted by die Deutsche Forschungsgemeinschaft (SFB 318) and by the Commission of the European Community under the Mast II programme (Contract MAS II–CT 93-0057).

References

Aagaard, K., Coachman, L. K. & Carmack, E. C. 1981 On the halocline of the Arctic Ocean. *Deep-Sea Res.* **28**, 529–545.

Aagaard, K., Swift, J. H. & Carmack, E. C. 1985 Thermohaline circulation in the Arctic Mediterranean Seas. *J. Geophys. Res.* **90**, 4833–4846.

Anderson, L. G., Björk, G., Holby, O., Jones, E. P., Kattner, G., Koltermann, K.-P., Liljeblad, B., Lindegren, R., Rudels, B. & Swift, J. H. 1994 water masses and circulation in the Eurasian Basin: results from the Oden 91 expedition. *J. Geophys. Res.* **99**, 3273–3283.

Bourke, R. H., Weigel, A. M. & Paquette, R. G. 1988 The westward turning branch of the West Spitsbergen Current. *J. Geophys. Res.* **93**, 14065–14077.

Backhaus, J. O. 1995 Proseßstudien zur Ozeanischen Konvektion. Hamburg: Habitulationsabhandlung der Universität Hamburg.

Carmack, E. C. 1986 Circulation and mixing in ice covered waters. In *The geophysics of sea ice* (ed. N. Untersteiner), pp. 641–712. New York: Plenum.

Carmack, E. C. 1990 Large-scale physical oceanography of polar oceans. In *Polar oceanography*, part A (ed. W. O. Smith Jr), pp. 171–212. San Diego, CA: Academic Press.

Clarke, R. A., Swift, J. H., Reid, J. L. & Koltermann, K.-P. 1990 The formation of Greenland Sea Deep water: double-diffusion or deep convection? *Deep-Sea Res.* **37**, 687–715.

Coachman, L. K. & Aagaard, K. 1974 Physical Oceanography of the Arctic and Sub-Arctic Seas. In *Marine Geology and Oceanography of the Arctic Ocean* (ed. Y. Herman), pp. 1–72. New York: Springer.

Dickson, R. R., Meincke, J., Malmberg, S.-A. & Lee, A. J. 1988 The 'great salinity anomaly' in the northern North Atlantic 1968–1982. *Prog. Oceanog.* **20**, 103–151.

Garwood, R. W. 1991 Enhancements to deep turbulent entrainment. In *Deep convection and deep water formation in the oceans* (ed. P. C. Chu & J. C. Gascard), pp. 197–213. Amsterdam: Elsevier.

Garwood, R. W., Isakari, S. M. & Gallacher, P. C. 1995 Thermobaric Convection. In *The polar oceans and their role in shaping the global climate* (ed. O. M. Johannessen, R. D. Muench & J. E. Overland), pp. 199–201. Washington, DC: American Geophysical Union.

GSP group 1990 Greenland Sea Project: a venture toward improved understanding of the oceans' role in climate. *Eos, Wash.* **71**, 750–751 & 754–755.

Houssais, M.-N. & Hibler III, W. D. 1993 Importance of convective mixing in seasonal ice margin simulations. *J. Geophys. Res.* **98**, 16427–16448.

Jones, E. P., Nelson, D. M. & Treguer, P. 1990 Chemical Oceanography. In *Polar oceanography*, part B (ed. W. O. Smith Jr), pp. 407–476. San Diego, CA: Academic Press.

Jones, E. P., Rudels, B. & Anderson, L. G. 1995 Deep waters of the Arctic Ocean: origins and circulation. *Deep-Sea Res.* (In the press.)

Jones, H. & Marshall, J. 1993 Convection with rotation in a neutral ocean: a study of open-ocean deep convection. *J. Phys. Ocean.* **23**, 1009–1039.

Kiilerich, A. 1945 On the hydrography of the Greenland Sea. *Medd. om Grönland* **144** (2), 63.

Loeng, H., Ozhigin, V., Ådlandsvik, B. & Sagen, H. 1993 Current Measurements in the northeastern Barents Sea. *ICES C. M. 1993/C* **41**. Hydrography Committee.

McCartney, M. S. 1982 The subtropical recirculation of mode waters *J. Mar. Res.* **40** (suppl.), 427–464.

McCartney, M. S. & Talley, L. D. 1984 Warm-to-cold water conversions in the northern North Atlantic Ocean. *J. Phys. Oceanogr.* **14**, 922–935.

Meincke, J., Jonsson, S. & Swift, J. H. 1992 Variability of convective conditions in the Greenland Sea. *ICES mar. Sci. Symp.* **195**, 32–39.

Meincke, J. & Rudels, B. 1995 Greenland Sea deep water: a balance between advection and convection. *Proc. ACSYS Conf., Göteborg, 7–10 November 1994, WMO Rep.* (In the press.)

Muench, R. D. 1990 Mesoscale phenomena in the polar oceans. In *Polar Oceanography*, part A (ed. J. O. Smith Jr), pp. 223–285. San Diego, CA: Academic Press.

Quadfasel, D. & Meincke, J. 1987 Note on the thermal structure of the Greenland Sea gyres. *Deep-Sea Res.* **34**, 1883–1888.

Quadfasel, D., Sy, A. & Rudels, B. 1993 A ship of opportunity section to the North Pole: upper ocean temperature observations. *Deep-Sea Res.* **40**, 777–789.

Rudels, B. 1986 The Θ–S relations in the northern seas: implications for the deep circulation. *Polar Res.* **4**, 133–159.

Rudels, B. 1987 On the mass balance of the Polar Ocean, with special emphasis on the Fram Strait. *Norsk Polarinstitutt Skrifter* **188**, 53.

Rudels, B. 1990 Haline convection in the Greenland Sea. *Deep-Sea Res.* **37**, 1491–1511.

Rudels, B., Quadfasel, D., Friedrich, H. & Houssais, M.-N. 1989 Greenland Sea convection in the winter of 1987–1988. *J. Geophys. Res.* **94**, 3223–3227.

Rudels, B. & Quadfasel, D. 1991 Convection and deep water formation in the Arctic Ocean–Greenland Sea System. *J. Mar. Syst.* **2**, 435–450.

Rudels, B., Meincke, J., Friedrich, H. & Schulze, K. 1993 Greenland Sea deep water: a report on the 1993 winter and spring cruises by RVs Polarstern and Valdivia. *ICES C. M. 1993/C* **59**. Hydrography Committee.

Rudels, B., Jones, E. P., Anderson, L. G. & Kattner, G. 1994 On the intermediate depth waters of the Arctic Ocean. In *The role of the polar oceans in shaping the global climate* (ed. O. M. Johannessen, R. D. Muench & J. E. Overland), pp. 33–46. Washington, DC: American Geophysical Union.

Rudels, B., Anderson, L. G. & Jones, E. P. 1995a Winter convection and seasonal sea-ice melt above the Arctic Ocean thermocline. *Proc. ACSYS Conf. Göteborg 7–10 Nov. 1994, WMO Rep.* (In the press).

Rudels, B., Friedrich, H., Hainbucher, D. & Lohmann, G. 1995b High latitude mixed layer in winter. *Deep-Sea Res.* (Submitted.)

Schauer, U., Rudels, B., Muench, R. D. & Timokhov, L. 1995 Circulation and water mass modifications along the Nansen Basin slope. *Proc. ACSYS Conf. Göteborg 7–10 Nov. 1994, WMO Rep.* (In the press.)

Schott, F., Visbeck, M. & Fisher, J. 1993 Observations of vertical currents and convection in the central Greenland Sea during the winter of 1988–1989. *J. Geophys. Res.* **98**, 14402–14421.

Strass, V. H., Fahrbach, E., Shauer, U. & Sellmann, L. 1993 Formation of Denmark Strait overflow water by mixing in the East Greenland Current. *J. Geophys. Res.* **98**, 6907–6919.

Swift, J. H., Aagaard, K. & Malmberg, S.-A. 1980 The contribution of the Denmark Strait overflow to the deep North Atlantic. *Deep-Sea Res.* **27**, 29–42.

Swift, J. H. & Aagaard, K. 1981 Seasonal transitions and water mass formation in the Icelandic and Greenland Seas. *Deep-Sea Res.* **28**, 1107–1129.

Walin, G. 1993 On the formation of ice on deep weakly stratified water. *Tellus* **45A**, 143–157.

Worthington, L. V. 1970 The Norwegian Sea as a mediterranean basin. *Deep-Sea Res.* **17**, 77–84.

Arctic sea ice extent and thickness

By P. Wadhams

Scott Polar Research Institute, University of Cambridge, Cambridge CB2 1ER, UK

Current knowledge on Arctic sea ice extent and thickness variability is reviewed, and we examine whether measurements to date provide evidence for the impact of climate change. The total Arctic ice extent has shown a small but significant reduction of $(2.1 \pm 0.9)\%$ during the period 1978–87, after apparently increasing from a lower level in the early 1970s. However, open water within the pack ice limit has also diminished, so that the reduction of sea ice area is only $(1.8 \pm 1.2)\%$. This stability conceals large interannual variations and trends in individual regions of the Arctic Ocean and sub-Arctic seas, which are out of phase with one another and so have little net impact on the overall hemispheric ice extent. The maximum annual global extent (occurring during the Antarctic winter) shows a more significant decrease of 5% during 1972–87. Ice thickness distribution has been measured by submarine sonar profiling, moored upward sonars, airborne laser profilometry, airborne electromagnetic techniques and drilling. Promising new techniques include: sonar mounted on an AUV or neutrally buoyant float; acoustic tomography or thermometry; and inference from a combination of microwave sensors. In relation to climate change, the most useful measurement has been repeated submarine sonar profiling under identical parts of the Arctic, which offers some evidence of a decline in mean ice thickness in the 1980s compared to the 1970s. The link between mean ice thickness and climatic warming is complex because of the effects of dynamics and deformation. Only fast ice responds primarily to air temperature changes and one can predict thinning of fast ice and extension of the open water season in fast ice areas. Another region of increasingly mild ice conditions is the central Greenland Sea where winter thermohaline convection is triggered by cyclic growth and melt of local young ice. In recent years convection to the bottom has slowed or ceased, possibly related to moderation of ice conditions.

---

1. Introduction

In this review we discuss evidence for climate-related changes in Arctic sea ice extent and thickness, and for ice-related changes in convection in the Greenland Sea.

The most important property of Arctic sea ice which affects its response to climate change is the fact that it is in motion, driven mainly by wind stress. Ice opens up under divergent stress, creating leads which rapidly refreeze in winter. Subsequent convergence or shear causes the leads to be crushed to create pressure ridges, semi-permanent features of the ice cover which contain about half of the total Arctic ice volume (Wadhams 1981). The wide distribution of thickness produced by mechanical processes results in ocean-atmosphere heat and moisture fluxes which are highly time- and space-dependent. A simple warming will not necessarily cause the ice cover as a

101

whole to become either thinner or less in extent; one has to consider several feedbacks associated with ice deformation and thickness variations.

Palaeoclimatic evidence suggests that some 18 000 years BP, during the most recent glaciation, sea ice extended down to the latitude of Spain and New England (Thiede, this volume). Thus, in contrast to the stable Antarctic Ice Sheet, sea ice does appear to be relatively unstable and responsive to climate change.

In §2 we consider evidence for sea ice extent variability, then in §3 we deal with the thickness distribution and how it varies. In §4 we consider how changes in Greenland Sea ice may be related to a change in the rate of winter convection, and in §5 we consider the mechanisms by which the extent and thickness of pack ice may respond to climate change.

2. Sea ice extent

(a) Importance in global change

Sea ice extent is defined as the ocean surface area enclosed by the ice edge. Within that boundary ice concentration may be less than unity and locally may even be zero, as in the case of persistent winter or spring polynyas such as the Northeast Water off northeastern Greenland or the North Water in Smith Sound. Therefore ice area is less than ice extent.

Sea ice area is related to global change by the same albedo feedback loop as snow cover on land. The albedo of sea ice ranges from 0.9 in winter after fresh snowfall, to 0.4–0.55 in summer when the surface is partly covered by meltwater pools. Since the albedo of open water is about 0.10, a decline in ice area will allow the absorption of more incoming short-wave radiation by the ocean and hence will feed back positively on surface temperature.

A further climatically important effect related to sea ice concentration is that most ocean-atmosphere heat loss occurs through open water or thin ice (Maykut 1986). In the case of open water the effect is not linear, in that when the concentration drops below $\frac{4}{10}$ the heat flux is almost as high as in open water (Worby & Allison 1991). Further, in summer, melting is enhanced in an open icefield because of the large floe perimeter per unit area.

(b) Measurement

The high contrast between ice and open water in both the visible and microwave spectral bands means that the advent of satellite observations permitted ice extent to be mapped on a worldwide basis. The first studies used visible or thermal infra-red imagery from radiometers on weather satellites, such as the AVHRR (advanced very high-resolution radiometer) on the NOAA satellites. Estimates of Arctic sea ice extent on an annually averaged basis from 1973 to 1990, for instance, were given in IPCC (1992, figure C13). The inability of such radiometers to see through cloud is a cause of inadequacy in the datasets, extending also to published ice charts which are usually constructed from a range of satellite data. It is interesting, however, that the general trend of ice extent in the 1992 IPCC chart (a Northern Hemisphere recovery in 1975–76 from an initially low value, followed by a slow subsequent decline) was reproduced successfully in predictions of total ice volume in the Arctic generated by Flato (1995) using a model which used daily varying geostrophic winds and monthly varying surface air temperatures to drive the ice.

A more effective means of measuring sea ice area and extent is by the use of satellite-borne passive microwave radiometers. These give all-weather day-and-night coverage of the polar regions apart from a small zone around the Poles. Algorithms relate microwave emissivity to ice concentration for single-frequency sensors, and yield additionally the fraction of older (multi-year) ice present in the case of multi-frequency sensors. Passive microwave data are available from 1973 to 1976 from the single-frequency ESMR (electrically scanning microwave radiometer), replaced from 1978 by the SMMR (scanning multichannel microwave radiometer) and in 1987 by the SSM/I (special sensor microwave/imager). Results from ESMR and SMMR have been summarized in atlases (Parkinson *et al.* 1987; Gloersen *et al.* 1992), which also discuss the method of data reduction and ice parameter extraction.

(*c*) *Evidence for changes in extent*

(i) *Hemispheric changes*

Ice extent data from these satellites demonstrated regional positive and negative trends in extent (Parkinson & Cavalieri 1989), with a slight negative overall trend for 1979–86 as compared to a slight positive trend for 1973–6. In a further analysis of the same data, Gloersen & Campbell (1991) corrected for instrumental drift and orbital errors, chose a 15% isopleth as the ice edge position, and used band-limited regression as the most appropriate statistical technique to search for long-term trends in a seasonally cyclic dataset. Their final result was a small but statistically significant reduction of $(2.1 \pm 0.9)\%$ in Arctic ice extent during the 8.8 years of SMMR (1978 87), with a confidence level of 96.5%, and a decrease in open water area within the pack limits of $(3.5 \pm 2.0)\%$ over the same period (confidence level 93.5%). The actual area of Arctic ice cover thus showed a less significant negative trend of $(1.8 \pm 1.2)\%$ with a significance of 88.5% (figure 1). No significant trend occurred in the Antarctic.

Present efforts are devoted to checking this relationship by extending the dataset backwards in time to include ESMR and forward to include SSM/I, but there are difficulties in achieving compatibility between algorithm products from different sensors.

In an earlier analysis (1988) Gloersen and Campbell had found that if Arctic and Antarctic sea ice extent are combined, the resulting global curve shows a significant negative trend in its peak value, amounting to 5% in 13 years. For this study the data were uncorrected and so the trend may be an over-estimate.

Even when ESMR and SSM/I data to the present (1994) can be included, the maximum duration of passive microwave time series will only be 22 years. As with all climatological datasets of limited duration, there is the possibility that an apparent trend detected over this time interval is actually part of a longer-term cycle.

(ii) *Regional changes*

A number of studies on regional changes in the Arctic have shown trends of decadal length (Walsh & Johnson 1979; Smirnov 1980; Mysak & Manak 1989; Parkinson & Cavalieri 1989). The SMMR atlas (Gloersen *et al.* 1992) divided the Arctic into nine regions, each of which was analysed separately for ice extent; ice area; open water area within the pack; and, in some cases during winter, multi-year fraction. Each region showed large interannual variations in each of these quantities, but with some out-of-phase effects between regions. The trends involved were analysed quantitatively by Parkinson and Cavalieri (1989) and again by Gloersen *et al.* (1992, p. 205). They found negative trends in the Sea of Okhotsk ($15\,000\ \mathrm{km^2\ a^{-1}}$), Kara/Barents Seas ($28\,000\ \mathrm{km^2\ a^{-1}}$) and Greenland Sea ($5000\ \mathrm{km^2\ a^{-1}}$); and generally smaller positive

P. Wadhams

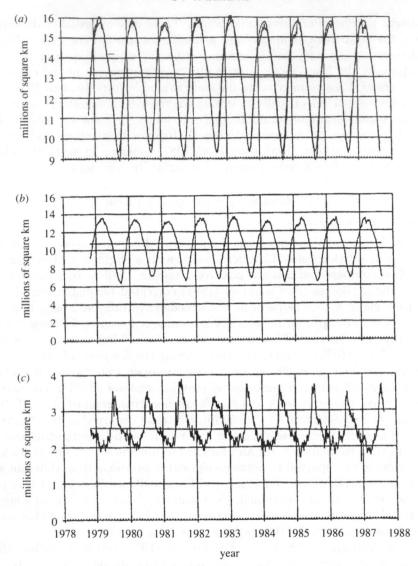

Figure 1. An 8.8-year cycle of Arctic sea ice (*a*) extent and (*b*) area, and (*c*) the open water area within the pack, together with trend lines. (After Gloersen *et al.* 1992.)

trends in Baffin Bay/Davis Strait (19 000 km² a⁻¹), Hudson Bay, the Bering Sea, the Gulf of St. Lawrence and the central Arctic Basin (all 4–6000 km² a⁻¹), the overall trend being a slightly negative one of 8000 km² a⁻¹.

Attempts have been made to associate periodicities in ice severity with periodic phenomena occurring in other parts of the world. Ono (1993) reported an apparent correlation between ice area in the Baffin Bay – Labrador Sea region as analysed by Mysak & Manak (1989), and indices of severity of the El Niño Southern Oscillation (ENSO). Figure 2 shows the Niño-3 sea surface temperature anomaly (anomaly in mean SST in the quadrangle 4° N, 4° S, 150° W, 90° W, a positive value indicating an El Niño year); the southern oscillation index (difference in sea level atmospheric pressure between Tahiti and Darwin, a negative value indicating atmospheric forcing linked to El Niño) and the ice area anomaly. There is an apparent correlation

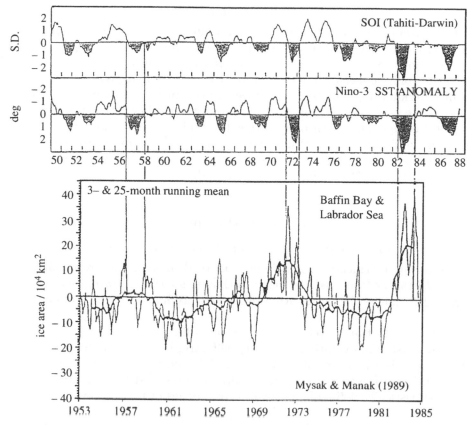

Figure 2. Ice extent anomaly in the Baffin Bay–Labrador Sea sector of the Arctic, compared with two indicators of the ENSO cycle. (After Ono 1993.)

involving three major El Niño events, but a causal link between tropical Pacific and northwestern Atlantic processes remains obscure. A more direct link might be expected with Bering Sea ice extent, and indeed if we compare figure 2 with fig. 3.4.2 of Gloersen *et al.* (1992) we find that El Niño events of 1982–3 and 1987 correspond with anomalously low winter ice extents in the Bering Sea. It is premature to comment on whether these correlations indicate association.

(iii) *Changes in the length of the ice season*

Using the 1979–86 SMMR dataset, Parkinson (1992) studied the length of the ice season, defined as the number of days in a calendar year during which a given grid element was covered by ice of concentration at least 30%. A least squares fit to a linear trend was obtained for each grid element. The results showed a consistent, spatially coherent pattern. During the 8-year period the season became significantly shorter off the north coast of Russia, and in the Greenland Sea, the Barents Sea and the Sea of Okhotsk, by often as much as 8 d a^{-1}, and significantly longer in the Gulf of St Lawrence, Labrador Sea, Hudson Bay, Beaufort Sea and eastern Bering Sea, by up to 16 d a^{-1}. This reinforced conclusions reached in the regional trend analysis, and demonstrated how these trends actually represent a shift in ice climatology during the SMMR period towards greater severity in the North American Arctic and greater mildness on the Eurasian side.

(d) Summary

The total Arctic ice extent has shown a small but statistically significant reduction of $(2.1 \pm 0.9)\%$ during the period 1978–87. Open water within the pack ice limits has also diminished, so that the reduction of ice area is only $(1.8 \pm 1.2)\%$. The underlying stability conceals large interannual variations and possible longer period trends in individual regions, some of which are completely isolated from an Arctic Ocean source (e.g. Sea of Okhotsk, Gulf of St. Lawrence). These trends have sufficient out-of-phase components with one another to result in only a minor overall trend in the hemispheric ice extent. When Arctic and Antarctic data are combined, the maximum annual global extent (occurring during the Antarctic winter) shows a more significant decrease of 5% during the same period.

3. Sea ice thickness

(a) The importance of ice thickness distribution

In considering ice thickness, it is important to deal with the probability density function $g(h)$ rather than just the mean thickness, for the following reasons:

(i) $g(h)$ determines the ocean-atmosphere heat exchange, with thin ice dominating;

(ii) together with the ice velocity, it gives mass flux;

(iii) its downstream evolution gives the melt rate, i.e. the fresh water flux;

(iv) its shape is a measure of the degree of deformation of the ice cover;

(v) if multi-year fraction is also known, $g(h)$ can be used to estimate ice strength and other statistically definable mechanical properties of the ice cover;

(vi) its variability is a test of model outputs;

(vii) its long term trend indicates the climatic response; and

(viii) in the central Greenland Sea the thickness achieved by young ice is associated with the magnitude of winter convection.

In addition to $g(h)$ it is also valuable to measure the ice bottom shape, requiring under-ice profiling rather than simply sampling the thickness at fixed intervals as is done with moored sonar. The advantages of this procedure are:

(i) shape is a determining factor for the aerodynamic and hydrodynamic drag coefficients;

(ii) the deepest pressure ridges are responsible for generating internal waves which may cause a significant internal wave drag;

(iii) seabed scour by the deepest ridges defines the fast ice limit on shelves and the extent of the stamukhi zone (Reimnitz et al. 1994);

(iv) ridges are an important component in the force exerted by an icefield on offshore structures;

(v) the scattering of underwater sound by ridges defines the range to which acoustic transmission can be accomplished, since upward refraction leads to repeated surface reflection; and

(vi) ridged ice provides additional habitats for sea ice biota.

The ice thickness distribution evolves with time through three processes: thermodynamic growth (or decay) which causes thin ice to grow thicker and thick ice to ablate; divergence of the ice cover, a source of open water and a sink of ice-covered area; and ridge formation, which moves ice from thin to thick categories in a way which is parameterized in models by a so-called ice thickness redistributor.

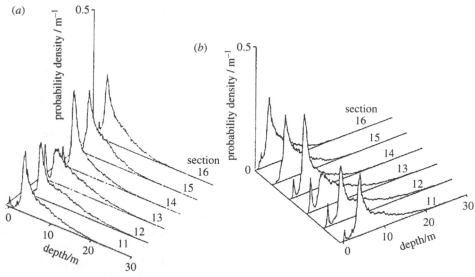

Figure 3. Some typical probability density functions of ice draft derived from successive 100 km submarine sonar profiles in the Arctic Basin at 85–90° N, 70° W. The two projections (*a*) and (*b*) emphasize the thick and thin ice respectively. (After Wadhams 1981.)

A typical ice thickness distribution (figure 3) has the following features: one or more peaks in the range 0–1.5 m due to ice in refrozen lead systems of different ages; a peak at about 2 m and another at about 3 m (sometimes the two are merged) representing undeformed first- and multi-year ice; and a tail, representing ridged ice, which shows an excellent fit to a negative exponential distribution (Wadhams 1981). If one views ridge-building as a Gibbs process, in which the work done translates into the potential energy acquired by the deformed ice, then a negative exponential form is the expected shape for this part of $g(h)$ (B. Kerman 1994, personal communication). The thickness distribution has been approximated by forms such as a gamma function (Goff 1995) or a sum of lognormals (Hughes 1991) for stochastic simulation purposes.

The ice surface topography has other statistical characteristics whose variability may have climatic implications. Pressure ridge keel depths and sail heights are observed to fit a negative exponential distribution (Wadhams 1981); pressure ridge spacings fit a log-normal (Wadhams & Davy 1986); and lead widths fit a power law of exponent about −1.45 for leads less than 100 m wide and −2.50 for wider leads (Wadhams 1992).

(*b*) How ice thickness is measured

(i) Present techniques

Methods of measuring ice thickness in the Arctic have been discussed by Wadhams (1994*a*) and Wadhams & Comiso (1992). Five methods are in common use. In decreasing order of total data quantity they are: submarine sonar profiling; moored upward sonar; airborne laser profilometry; airborne electromagnetic techniques; and drilling.

Submarine sonar can obtain synoptic data on thickness and under-ice topography rapidly and accurately, and most information on the distribution of $g(h)$ over the Arctic comes from upward-looking submarine sonar profiles. The addition of sidescan or swath sounding sonar adds information on ice type and two-dimensional bottom topography (Wadhams 1988). However, repeated profiling over identical tracks to test for climate-related trends, or systematic grid surveys, are both difficult with military submarines since ice profiling is an addendum to their operational tasks.

Moored upward sonar gives long-term information from a single point. It is invaluable for assessing the time variation of ice flux through critical regions such as Fram Strait. However, the cost and difficulty of deployment and recovery preclude its general use on a systematic measurement grid over the central Arctic Basin.

Airborne laser profilometry yields a freeboard distribution which can be converted to draft distribution if the mean density of ice plus overlying snow is known (Wadhams *et al.* 1991). Seasonal and regional validation of snow depths is needed before this otherwise rapid and efficient technique can be used for basinwide surveys.

Airborne electromagnetic techniques involve generating and sensing eddy currents under ice by VLF (10–50 kHz) EM induction from a coil towed behind a helicopter, with a simultaneous laser to give range to the snow surface. The technique has been reviewed by Rossiter and Holladay (1994). The wide footprint involves loss of resolution of individual ridges and a need to fly very low. Recently a system has been mounted in a Twin Otter by the Geological Survey of Finland, giving greater range.

Drilling is the most accurate, but slowest, technique, the ultimate validation for all others.

(ii) *Future techniques*

Synoptic, repeated ice thickness surveys over the Arctic Basin are essential for detecting a climatic signal in ice thickness. Novel techniques may be needed, such as:

(i) Sonar surveys by a dedicated submarine, through temporary reassignment from the military, full conversion to a civilian research vessel (the so-called 'white submarine' concept), or by use of a commercially designed small submarine.

(ii) Mounting sonars on an unmanned vehicle. For short-range surveys this could be a cable-controlled ROV (remotely operated vehicle), and for mesoscale and basin-wide surveys an AUV (autonomous underwater vehicle). Vehicles with 300 km range are in service (Tonge 1992), and vehicles of longer range are under development.

(iii) Mounting sonar on a neutrally buoyant float, to construct $g(h)$ over, say, a week's drift, the data being transmitted acoustically to a readout station. This requires only a modest extension of the existing technology of under-ice SOFAR and RAFOS floats.

(iv) The use of acoustic techniques. It has been shown that travel time changes for an acoustic path are reduced by the presence of an ice cover, by an amount approximately proportional to the ice thickness (Guoliang & Wadhams 1989; Jin *et al.* 1993). In long range acoustic propagation experiments this can be used to give a single mean value for ice thickness along a path. In spring 1994 the first transmission across the Arctic Basin from Svalbard to the Beaufort Sea was successfully accomplished.

(v) Increased efforts to obtain empirical correlations between ice thickness and the output of satellite sensors such as passive and active microwave or altimeter. Already a positive correlation between SAR backscatter and ice thickness has been demonstrated (Wadhams & Comiso 1992). Further advance requires extensive validation.

(vi) Deriving ice thickness as a by-product of another measurement. For instance, long distance swell propagation in ice is subject to a slow attenuation due to creep (Wadhams 1973). The attenuation rate is frequency- and thickness-dependent. In principle one could obtain spectra of flexure from sets of strainmeters or tiltmeters across the Arctic, and derive a mean value for ice thickness from the attenuation rate.

An Arctic mapping strategy might be to combine repeated under-ice sonar profiles over a grid covering the Basin, with moored upward sonar measurements spanning key choke points for ice transport, i.e. Fram Strait, the Svalbard–Franz Josef Land gap and a small number of specimen points within the Trans Polar Drift Stream and Beaufort Gyre. Sonar moorings could be combined with current meters and sediment traps. If submarines are not available for the task, an alternative would be a combination of long-range AUVs, airborne laser surveys and (for regional measurements) airborne EM. At the same time, the validation of satellite sensors can be developed. Ice thickness mapping is being addressed by the Arctic Climate System Study (ACSYS) of WCRP (WCRP 1994).

(c) Present knowledge of Arctic ice thickness

Submarine sonar profiling shows that over the Arctic Basin there is a gradation in mean ice thickness from the Russian Arctic, across the Pole and towards the coasts of north Greenland and the Canadian Arctic Archipelago, where the highest mean thicknesses of some 7–8 m are observed (LeSchack 1980; Wadhams 1981, 1992; Bourke & McLaren 1992). These overall variations are in accord with the predictions of numerical models (Hibler 1979, 1980) which take account of ice dynamics and deformation as well as thermodynamics. Large scale maps of the general distribution of mean thickness have been generated by LeSchack (1980), Bourke & Garrett (1987) (figure 4) and Bourke & McLaren (1992). The data used for the Bourke and Garrett map did not include open water, so these maps are over-estimates of the mean ice draft. The update by Bourke & McLaren (1992) included contour maps of standard deviation of draft, and mean pressure ridge frequencies and drafts for summer and winter, based on 12 submarine cruises.

Measurements in sub-Arctic seas show that the ice in Baffin Bay is largely thin first-year ice with a modal thickness of 0.5–1.5 m (Wadhams et al. 1985). In the southern Greenland Sea the ice, although composed largely of partly melted multi-year ice, also has a modal thickness of about 1 m (Vinje 1989; Wadhams 1992), with the decay rate in thickness from Fram Strait giving a measure of the fresh water input to the Greenland Sea at different latitudes.

(d) Evidence for changes

In order to understand whether, and how, the thickness of sea ice in the Arctic is responding to climate change it is necessary to measure $g(h)$ repetitively on a synoptic scale, preferably using the same equipment. Data available at present offer only indications that climate-related change may be occurring.

McLaren (1989) compared data from two US submarine transects of the Arctic Ocean in August 1958 and August 1970, running from Bering Strait to Fram Strait via the North Pole. He found no major changes in the Eurasian Basin and North Pole area, but significantly milder conditions in the Canada Basin in 1970. The difference is possibly due to anomalous cyclonic activity as observed in the region in some summers (Serreze et al. 1989). Also, since August is the month of greatest ice retreat in the Beaufort Sea, the difference may be due to differences between the ice edge positions in the Chukchi and southern Beaufort Seas during the respective summers.

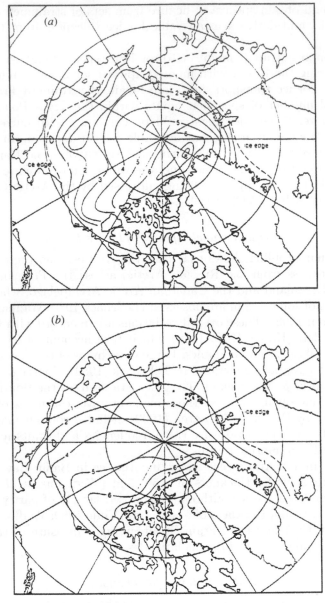

Figure 4. Contour maps of estimated climatology of mean ice thickness for (*a*) summer (*b*) winter in the Arctic Basin, neglecting open water. (After Bourke & Garrett 1987.)

An unusually open southern Beaufort Sea would lead to more open conditions within the pack itself.

Wadhams (1989) compared mean ice drafts for a region of the Eurasian Basin north of Fram Strait, from British cruises carried out in October 1976, April–May 1979 and June–July 1985, all using similar sonar equipment. A box extending from 83°30′ N to 84°30′ N and from 0° to 10° E had an especially high track density from the three cruises (400 km in 1976, 400 km in 1979 and 1800 km in 1985). It is distant from land boundaries, and is representative of the Trans Polar Drift Stream prior to

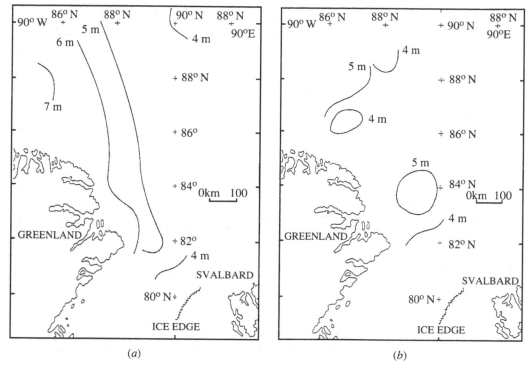

Figure 5. Contour maps of mean ice drafts from Eurasian Basin: (*a*) October 1976 and (*b*) May 1987. (After Wadhams 1990*a*.)

the acceleration and narrowing which occur in Fram Strait. The mean drafts from the three cruises were remarkably similar: 4.60 m in 1976; 4.75 m in 1979; and 4.85 m in 1985, despite being recorded in different seasons as well as different years.

Using newer data, Wadhams (1990*a*) compared data from a triangular region of 300 000 km^2 extending from north of Greenland to the North Pole, recorded in October 1976 and May 1987. Mean drafts were contoured to give the maps shown in figure 5. Over the whole area there was a decrease of 15% in mean draft from 5.34 m in 1976 to 4.55 m in 1987. The decrease was concentrated in the region south of 88° N and between 30° and 50° W. An analysis of the shape of the distributions showed that in 1987 there was more ice present in the form of young ice in refrozen leads (stretches of ice with draft less than 1 m) and as first-year ice (draft less than 2 m). There was less multi-year ice (interpreted as ice 2–5 m thick) and less ridging (ice more than 5 m thick) in 1987. The main contribution to the loss of volume was thus the replacement of multi-year and ridged ice by young and first-year ice.

To determine the mechanism, tracks of drifting buoys from the Arctic Ocean Buoy Program (Colony *et al.* 1991) were examined. Four buoys were in the region during the months prior to the 1987 cruise. Three of the buoys, located in the Beaufort Gyre, remained almost stationary during January – May 1987, while the fourth, in the Trans Polar Drift Stream, moved towards Fram Strait at a mean speed of 2 km d^{-1}. The result of this anomalous halting of the motion of part of the Beaufort Gyre should be a divergence within the experimental region, leading to the opening up of the pack and the creation of young and first-year ice. Thus an ice motion anomaly rather than, or as well as, an ice growth anomaly could cause the observed

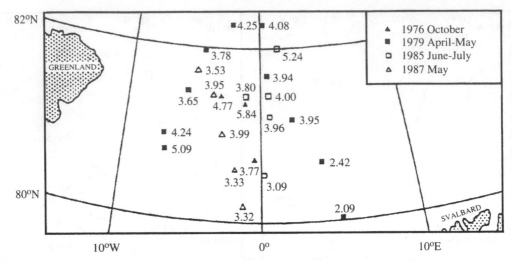

Figure 6. Comparisons of mean ice drafts measured in the region north of Fram Strait (after Wadhams 1992). Each position is the centroid of a 50 km section of submarine profile.

decrease in mean ice draft. This supposition is supported by modelling results from Chapman & Walsh (1993), who ran the Hibler ice model using daily wind forcing and monthly thermodynamic forcing, both varying interannually. They found that the simulated thicknesses showed negative anomalies in this region in May 1987 and positive anomalies in October 1976. A build-up of mean ice draft towards Greenland appears in the model predictions of Hibler (1980) and in the tentative seasonal climatology of Bourke & Garrett (1987) and was previously thought to be a stable aspect of ice climatology. Clearly the ice cover, like the ocean, possesses a weather as well as a climate.

The 1987 dataset allowed a further regional comparison immediately north of Fram Strait, between 82° N and 80° N, where data from 1976 1979 1985 and 1987 are now available. This is a mixing zone between streams of old, deformed ice moving S into Fram Strait from the North Pole region and SE from the region north of Greenland; and a stream of younger, less deformed ice moving SW from the seas north of Russia. Figure 6 shows all available mean drafts from 50 km sections (from Wadhams 1981, 1983, 1989 and current analyses). There is good consistency among the data, regardless of year or season; fluctuations appear to be random in character, and where results from different experiments lie close to one another, the mean drafts are usually similar. Only the 1976 data points appear anomalously thick.

McLaren *et al.* (1992) analysed 50 km and 100 km sections of ice profile centred on the North Pole from 6 cruises from 1977 to 1990. They found that the mean ice draft from 50 km sections in the late 1970s (1977, 1979) was 4.1 m (4.2, 4.0 m respectively), while the mean draft for the late 1980s was 3.45 m (2.8, 4.1, 3.3, 3.6 m for 1986, 1987, 1988, 1990). They showed using a t-test that the difference of 0.65 m (15%) between the means is significant only at the 20% level, i.e. non-significant. They also claimed that this showed that the 15% decline in mean drafts found by Wadhams (1990a) between 1976 and 1987 is similarly non-significant. The claim is invalid because the McLaren *et al.* (1992) comparison is between two datasets of total length 100 km and 200 km, while the Wadhams comparison is between two datasets of length 3900 km

(1976) and 2200 km (1987), giving a much greater statistical stability to the mean values.

An alternative way of interpreting the results of McLaren *et al.* (1992) suggests that they are more significant than their authors state. When mean ice draft is computed from a dataset of finite length it is subject to a statistical variability which is independent of variations in ice climatology and which is merely due to the fact that the profile is sampling only a finite number of ice floes, leads and pressure ridges. This variability may be termed the sampling error. There is no *a priori* way of estimating what the sampling error should be. The only valid method is an experimental one: to examine the statistical stability of sections taken from long ice profiles obtained within a homogeneous ice regime at a single instant. Wadhams *et al.* (1992) discussed two such experimental datasets and found them compatible. The first comprised 23 50 km sections obtained from a restricted area of the Beaufort Sea (Wadhams & Horne 1980) while the second comprised 18 100 km sections from a homogeneous part of the central Arctic Basin (Wadhams 1981). If we assume that the ice regime was homogeneous, then the standard error in the means of the 50- or 100 km sections is the sampling error which we seek. The results from the two experimental datasets were that the 50 km sections gave (3.67 ± 0.19) m as the overall mean draft (a 5.2% sampling error) while the 100 km sections gave (4.51 ± 0.18) m (a 3.9% sampling error). The difference in mean is due to the different ice regimes, but the percentage sampling errors are almost identical, given that the sampling length is double in the second case (if they were completely identical, a 5.2% error in 50 km would become a 3.7% error in 100 km). Assuming that a 3.9% sampling error per 100 km is indeed characteristic of Arctic sea ice, and applying it to the McLaren *et al.* (1992) data listed above, we see that on the hypothesis that all 1970s data and all 1980s data were drawn from only two populations the means and sampling errors would be (4.1 ± 0.16) m (1970s) and (3.45 ± 0.10) m (1980s). The appropriate test for the significance of the difference between the means is to compute the d statistic of the Fisher Behrens distribution (Campbell 1974), which gives $d = 3.444$, significant at better than the 1% level. The data presented therefore do show evidence of a difference between 1970s and 1980s data at the North Pole.

Further data supportive of the above are some recently analysed, hitherto unpublished, profiles from the region within 1° of the North Pole, obtained by the author during an earlier phase of the 1987 experiment. An analysis of the 10 50 km sections obtained gives a mean draft of 3.845 m with a standard error of 0.155 m, implying a 4.0% sampling error in 50 km of track. This is lower than the error observed in the previous two experiments, and when applied to the McLaren *et al.* data gives an even more significant difference between the mean thicknesses for the two decades.

(e) *Summary*

From limited data comparisons made to date, the following results were obtained on thickness variability:

(i) Ice in parts of the Trans-Polar Drift Stream which are not heavily influenced by a downstream land boundary show interseasonal and interannual consistency in mean thickness, especially at latitudes from 84°30′ N to 80° N and longitudes near 0°.

(ii) Ice near the land boundary of Greenland can show large variability in mean draft, notably a significant decline between 1976 and 1987, but this can be ascribed to a variable balance between pressure ridge formation through convergence and open water formation through divergence. Deeper knowledge of ice dynamics and more adequate data are needed to understand these changes fully.

(iii) Data obtained from the North Pole region in 1977–90 show evidence of a decline in mean ice thickness in the late 1980s relative to the late 1970s.

(iv) Ice in the Canada Basin in summer also shows variability in mean ice draft, but here there is a free boundary with ice-free marginal seas, permitting relaxation of the ice cover into a less concentrated state under certain wind conditions.

There is no conclusive evidence of progressive thinning of the sea ice cover, as would be caused by the impact of the greenhouse effect, but the data are suggestive enough to make more systematic data collection essential.

4. Ice variations and ocean convection

One region in which ice retreat and thinning may have already triggered an ocean response is the centre of the Greenland Sea gyre. South of the main gyre centre an ice tongue usually develops during winter, growing eastward from the main East Greenland ice edge in 72–74° N latitude and often curving round to the northeast until it reaches east of the prime meridian. It is called Odden, its curvature embracing a bay of open water known as Nordbukta. Odden forms mainly by local ice production on the cold surface water of the Jan Mayen Polar Current (the southern part of the Greenland Sea Gyre), largely in the form of frazil and pancake ice, since the intense wave field inhibits ice sheet formation. Frazil and pancake production implies high growth rates and high salt fluxes, possibly on a cyclic basis related to cold air outbreaks from Greenland. In recent models of winter convection in the central Greenland Sea (Rudels 1990; J. Backhaus, personal communincation), this periodic salt flux is responsible for triggering narrow convective plumes.

Field operations during the European Subpolar Ocean Programme (ESOP) in 1993–94 involved the direct study of convective plumes and water structure in the central gyre region, and simultaneous investigations of ice characteristics (Wadhams 1994b). In 1993 Odden developed as a tongue, later an island, of dense pancake and frazil ice, whose ability to grow and melt rapidly gave the Odden a rapidly changing shape on SSM/I images. The ice physics programme (Wadhams & Viehoff 1994; Wadhams et al. 1994) involved recovery of pancakes and frazil, with measurements of thickness and salinity. By correlating ice properties to assumed age, an estimate was made of ice volume and salinity changes involved in the winter fluctuations of Odden; the resulting salt fluxes (Wadhams et al. 1995) could then be compared to fluxes required to trigger convection in model studies.

If under global warming ice were to cease forming in Odden it may cause deep convection to cease. Already tracer studies (Schlosser et al. 1991) show a severe reduction in the renewal of the deep waters of the Greenland Sea by convection during the last decade. Convection reached only 1000 m in 1993 and 400 m in 1994 when Odden did not form at all (D. Quadfasel, personal communication). If deep convection were to cease it would have a positive feedback effect on global warming, since the ability of the world ocean to sequester CO_2 through convection would be reduced (Rudels, this volume). A general weakening of the thermohaline circulation in the northern North Atlantic is predicted by Manabe & Stouffer (1994) based on the freshening of surface water due to melt of sea ice and ice sheets.

5. Climatic feedbacks and model predictions

(a) Extent

As surface air temperatures increase, it is legitimate to question how ice extent will vary. One test is to examine correlations between ice extent and past temperature changes. Chapman & Walsh (1993) showed that longitudinal patterns of sea ice variability in the Arctic during the periods 1961–75 and 1976–90 had a significant negative correlation with annual air temperature changes in the latitude zone 55–75° N, i.e. the latitude range corresponding approximately to the seasonal ice edge. Most transient general circulation models (GCMs) predict a retreat of sea ice: Manabe *et al.* (1990), for instance, predicted a significant northward retreat of the ice edge under a CO_2 doubling, with thinning throughout the ice cover but with the effect being greatest in the region 60–70° N, i.e. the sub-Arctic. Other recent models predict that eventually the ice cover will wholly disappear in summer, and will become essentially seasonal such as the Antarctic today.

(b) Fast ice thickness

The simplest effect of warming will be on the thickness of fast ice, which grows in fjords, bays and inlets in the Arctic, along the open coast in shallow water, and in channels of restricted dimensions. Here oceanic heat flux is negligible, and the ice thickness is determined by air temperature history modified by the thickness of the snow cover. Empirical relationships have been successfully developed relating thickness achieved to the degree-days of freezing since the beginning of winter (e.g. Bilello 1961). If the average daily air temperature increases by a known amount, the ultimate ice thickness will diminish by an amount which is easily extracted from these relationships, and the ice-free season will lengthen. Using this technique, Wadhams (1990b) predicted that in the Northwest Passage and Northern Sea Route an air temperature rise of 8 °C (equivalent to about a century of warming) will lead to the winter fast ice thickness declining from 1.8–2.5 m (depending on snow thickness) to 1.4–1.8 m and the ice-free season increasing from 41 to 100 days.

Even in this simple case, however, there is a feedback with snow thickness. If Arctic warming produces increased open water area and thus increased atmospheric water vapour content, it will lead to increased precipitation. Thicker snow cover decreases the growth rate of fast ice, as has been directly observed (Brown & Cote 1992), except if the thickness increases to the point where the snow does not all melt in summer, in which case the protection that it offers the ice surface from summer melt leads to a large increase in equilibrium ice thickness (Maykut & Untersteiner 1971).

(c) Moving pack ice thickness

In moving pack ice, thermodynamic growth and decay rates no longer determine the area-averaged mean thickness. Pressure ridge building causes a redistribution of ice from thinner to thicker categories, with the accompanying creation of open water areas. It also makes the ice cover as a whole more resistant to convergent than to divergent stresses, and this causes its motion field under wind stress to differ from that of the surface water. Thus the exchanges of heat, salt and momentum all differ from those that would occur in a fast ice cover.

The variable thickness has thermodynamic effects. It has been found that the overall area-averaged growth rate of ice is dominated (especially in autumn and early winter when much lead and ridge creation take place) by the small fraction of the sea surface occupied by ice less than 1 m thick (Hibler 1980). In fast ice, climatic warming increases sea-air heat transfer by reducing ice growth rates. However, over open leads a warming would decrease the sea-air heat transfer, so the area-averaged change in this quantity over moving ice (and hence its feedback effect on climatic change itself) depends on the change in the rate of creation of new lead area, which is itself a function of a change in the ice dynamics, either driving forces (wind field) or response (ice rheology).

The overall pattern of ice motion has other effects. In the Eurasian Basin the average surface ice drift pattern is a current (the Trans-Polar Drift Stream) which transports ice across the Basin, out through Fram Strait, and south via the East Greenland Current into the Greenland and Iceland Seas where it melts. A typical parcel of ice forms by freezing in the Basin, the latent heat being transferred to the atmosphere; is then transported southward (equivalent to a northward heat transport); and then when it melts in the Greenland or Iceland Sea it absorbs the latent heat required from the ocean. The net result is a heat transfer from the upper ocean in sub-Arctic seas into the atmosphere above the Arctic Basin. A change in area-averaged freezing rate in the Basin would thus cause a change of similar sign to the magnitude of this long-range heat transport. An identical argument applies to salt flux, which is positive into the upper ocean in ice growth areas and negative in melt areas. Thus salt is also transported northward via the southward ice drift. A relative increase in area-averaged melt would cause increased stabilization of the upper layer of polar surface water, and hence a reduction in heat flux by mixing across the pycnocline, while a relative increase in freezing would cause destabilization and possible overturning and convection.

Finally, Hibler (1989) has drawn attention to the role of ice deformation in reducing the sensitivity of ice thickness to global warming in areas of net convergence. The largest mean ice thicknesses in the Arctic – 7 m or more – occur off the Canadian Arctic Archipelago (Hibler 1979) where ice is driven towards a downstream land boundary. Here the mean ice thickness is determined by mechanical factors, largely the strength of the ice, which sets a limit to the amount of deformation by crushing that can occur. In this area the thickness is likely to be insensitive to atmospheric temperature changes. The main sensitivity would be to a change in the overall wind pattern over the Arctic.

Given the complexity of these interactions and feedbacks, it is not clear at present what the quantitative effect of an air temperature increase on the Arctic ice cover and upper ocean would be. Further sensitivity studies using coupled ocean-ice-atmosphere models are required.

For work described in this review the author is grateful for support from the Commission of the European Communities under contract MAS2-CT93-0057 of the MAST-II programme; the US Office of Naval Research, Arctic Program; the Defence Research Agency; and the Natural Environment Research Council.

References

Bilello, M. A. 1961 Formation, growth and decay of sea ice. *Arctic* **14**, 3–24.

Bourke, R. H. & Garrett, R. P. 1987 Sea ice thickness distribution in the Arctic Ocean. *Cold Regions Sci. Technol.* **13**, 259–280.

Bourke. R. H. & McLaren, A. S. 1992 Contour mapping of Arctic Basin ice draft and roughness parameters. *J. geophys. Res.* **97**, 17 715–17 728.

Brown, R. D. & Cote, P. 1992 Interannual variability of landfast ice thickness in the Canadian High Arctic 1950–1989. *Arctic* **45**, 273–284.

Campbell, R. C. 1974 *Statistics for Biologists*, 2nd edn. Cambridge University Press.

Chapman, W. L. & Walsh, J. E. 1993 Recent variations of sea ice and air temperature in high latitudes. *Bull. Am. Meteorol. Soc.* **74**, 33–47.

Colony, R. L., Rigor, I. & Runciman-Moore, K. 1991 A summary of observed ice motion and analyzed atmospheric pressure in the Arctic Basin 1979–1990. *Appl. Phys. Lab., Univ. Washington, Seattle, WA, Tech. Rep.* APL-UW TR9112.

Flato, G. M. 1995 Spatial and temporal variability of Arctic ice thickness. *Ann. Glaciol.* (In the press.)

Gloersen, P. & Campbell, W. J. 1988 Variations in the Arctic, Antarctic and global sea ice covers during 1978–1987 as observed with the Nimbus 7 Scanning Multichannel Microwave Radiometer. *J. Geophys. Res.* **93**, 10 666–10 674.

Gloersen, P. & Campbell, W. J. 1991 Recent variations in Arctic and Antarctic sea- ice covers. *Nature, Lond.* **352**, 33–36.

Gloersen, P., Campbell, W. J., Cavalieri, D. J., Comiso, J. C., Parkinson, C. L. & Zwally, H. J. 1992 *Arctic and Antarctic Sea Ice 1978–1987: satellite passive-microwave observations and analysis*. Washington, D.C.: NASA SP-511.

Goff, J. A. 1995 Quantitative analysis of sea ice draft. I. Methods for stochastic modeling. *J. geophys. Res.* **100**, 6993–7004.

Guoliang, J. & Wadhams, P. 1989 Travel time changes in a tomography array caused by a sea ice cover. *Prog. Oceanogr.* **22**, 249–275.

Hibler, W. D. III 1979 A dynamic thermodynamic sea ice model. *J. phys. Oceanogr.* **9**, 815–846.

Hibler, W. D. III 1980 Modeling a variable thickness sea ice cover. *Mon. Weather Rev.* **108**, 1943–1973.

Hibler, W. D. III 1989 Arctic ice-ocean dynamics. In *The Arctic seas: climatology, oceanography, geology, and biology* (ed. Y. Herman), pp. 47–91. New York: Van Nostrand Reinhold.

Hughes, B. A. 1991 On the use of lognormal statistics to simulate one- and two- dimensional under-ice draft profiles. *J. geophys. Res.* **96**, 22 101–22 111.

IPCC 1992 *Climate change 1992: the supplementary report to the IPCC scientific assessment* (ed. J. T. Houghton, B. A. Callander & S. K. Varney). Cambridge University Press.

Jin, Guoliang, Lynch, J. F., Pawlowicz, R., Wadhams, P. & Worcester, P. 1993 Effects of sea ice cover on acoustic ray travel times, with applications to the Greenland Sea Tomography Experiment. *J. Acoust. Soc. Am.* **94**, 1044–1056.

LeSchack, L. A. 1980 *Arctic Ocean sea ice statistics derived from the upward-looking sonar data recorded during five nuclear submarine cruises*, pp. 116–1111. Silver Spring, MD: LeSchack Associates Limited. (Technical Report.)

McLaren, A. S. 1989 The under-ice thickness distribution of the Arctic basin as recorded in 1958 and 1970. *J. geophys. Res.* **94**, 4971–4983.

McLaren, A. S., Walsh, J. E., Bourke, R. H., Weaver, R. L. & Wittman, W. 1992 Variability in sea-ice thickness over the North Pole from 1979 to 1990. *Nature, Lond.* **358**, 224–226.

Manabe, S. & Stouffer, R. J. 1994 Multiple-century response of a coupled ocean-atmosphere model to an increase of atmospheric carbon dioxide. *J. Climate* **7**, 5–23.

Manabe, S., Bryan, K. & Spelman, M. J. 1990 Transient response of a global ocean-atmosphere model to the doubling of atmospheric carbon dioxide. *J. phys. Oceanogr.* **20**, 722–749.

Martin, S. & Kauffman, P. 1981 A field and laboratory study of wave damping by grease ice. *J. Glaciol.* **27**, 283–313.

Maykut, G. A. 1986 The surface heat and mass balance. In *The geophysics of sea ice* (ed. N. Untersteiner), pp. 1163–1183. New York: Plenum.

Maykut, G. A. & Untersteiner, N. 1971 Some results from a time-dependent thermodynamic model of Arctic sea ice. *J. Geophys Res.* **76**, 1550–1575.

Mysak, L. A. & Manak, D. K. 1989 Arctic sea ice extent and anomalies 1953–84. *Atmos. Ocean.* **27**, 346–505.

Ono, N. 1993 ENSO-related phenomena in the Arctic. In *Proc. Joint Japanese–Norwegian Workshop on Arctic Research* (ed. S. Kokobun, N. Matuura, A. Egeland, T. Amundsen), pp. 46–49. Nagoya University.

Parkinson, C. L. 1992 Spatial patterns of increases and decreases in the length of the sea ice season in the north polar region 1979–1986. *J. geophys. Res.* **97**, 14377–14388.

Parkinson, C. L. & Cavalieri, D. J. 1989 Arctic sea ice 1983–1987: seasonal, regional and interannual variability. *J. Geophys. Res.* **94**, 14 199–14 523.

Parkinson, C. L., Comiso, J. C., Zwally, H. J., Cavalieri, D. J., Gloersen, P., Campbell, W. J. 1987 *Arctic sea ice 1973–1976: satellite passive-microwave observations*, NASA SP-489. Washington, DC: National Aeronautics and Space Administration.

Reimnitz, E., Dethleff D. & D. Nürnberg 1994 Contrasts in Arctic shelf sea-ice regimes and some implications: Beaufort Sea versus Laptev Sea. *Marine Geol.* **119**, 215–225.

Rossiter, J. R. & Holladay, J. S. 1994 Ice-thickness measurement. In *Remote sensing of sea ice and icebergs* (ed. S. Haykin, E. O. Lewis, R. K. Raney & J. R. Rossiter), pp. 141–176. New York: Wiley.

Rudels, B. 1990 Haline convection in the Greenland Sea. *Deep-Sea Res.* **37**, 1491–1511.

Schlosser, P., Bönisch, G., Rhein M. & Bayer, R. 1991 Reduction of deepwater formation in the Greenland Sea during the 1980s: evidence from tracer data. *Science, Wash.* **251**, 1054–1056.

Serreze, M. C., Barry, R. G. & McLaren, A. S. 1989 Summertime reversals of the Beaufort Gyre and its effects on ice concentration in the Canada Basin. *J. geophys. Res.* **94**, 10 955–10 970.

Smirnov, V. A. 1980 Opposition in the redistribution in the waters of the foreign Arctic. *Sov. Meteorol. Hydrol.* **3**, 73–77. (English translation.)

Tonge, A. M. 1992 An incremental approach to AUVs. In *Proc. Oceanol. Int. '92, Brighton, 10–13 March 1992*, vol. 1. London: Spearhead Exhibitions.

Vinje, T. E. 1989 An upward looking sonar ice draft series. In *Proc. 10th Int. Conf. Port Ocean Engng Arctic Cond.* (ed. K. B. E. Axelsson, L. A. Fransson), vol. 1, pp. 178–187. Luleå: Luleå University of Technology.

WCRP 1994 *Arctic climate system study (ACSYS): initial implementation plan*, report WCRP-85. Geneva: World Meteorological Organization.

Wadhams, P., 1973 Attenuation of swell by sea ice. *J. Geophys. Res.* **78**, 3552–3565.

Wadhams, P. 1981 Sea ice topography of the Arctic Ocean in the region 70° W to 25° E, *Phil. Trans. R. Soc. Lond.* A **302**, 45–85.

Wadhams, P. 1983 Sea ice thickness distribution in Fram Strait. *Nature, Lond.* **305**, 108–111.

Wadhams, P. 1988 Sidescan sonar imagery of Arctic sea ice. *Nature, Lond.* **333**, 161–164.

Wadhams, P. 1989 Sea-ice thickness in the Trans-Polar Drift Stream. *Rapp. P-v Reun Cons. Int. Explor. Mer* **188**, 59–65.

Wadhams, P. 1990*a* Evidence for thinning of the Arctic ice cover north of Greenland. *Nature, Lond.*, **345**, 795–797.

Wadhams, P. 1990*b* Sea ice and economic development in the Arctic Ocean – a glaciologist's experience. In *Arctic technology and economy: present situation and problems, future issues*, pp. 1–23. Paris: Bureau Veritas.

Wadhams P. 1992 Sea ice thickness distribution in the Greenland Sea and Eurasian Basin, May 1987. *J. geophys. Res.* **97**, 5331–5348.

Wadhams, P. 1994*a* Sea ice thickness changes and their relation to climate. In *The polar oceans and their role in shaping the global environment: the Nansen centennial volume* (ed. O. M. Johannessen, R. D. Muench & J. E. Overland), Geophys. Monograph 85, pp. 337–361. Washington: American Geophysical Union.

Wadhams, P. 1994*b* The European subpolar ocean programme (ESOP). In *Proc. Oceanol. Int. 1994, Brighton, March 8–11 1994*, vol. 1. London: Spearhead Exhibitions.

Wadhams, P. & Comiso, J. C. 1992 The ice thickness distribution inferred using remote sensing techniques. In *Microwave remote sensing of sea ice* (ed. F. Carsey), Geophysical Monograph 68, pp. 375–383. Washington: American Geophysical Union.

Wadhams, P. & Horne, R. J. 1980 An analysis of ice profiles obtained by submarine sonar in the Beaufort Sea. *J. Glaciol.* **25**, 401–424.

Wadhams, P. & Viehoff, T. 1994 The Odden ice tongue in the Greenland Sea: SAR imagery and field observations of its development in 1993. In *Proc. 2nd ERS-1 Symp. – Space at the Service of our Environment, Hamburg, 11–14 October 1993*, SP-361, vol. 1. Paris: European Space Agency.

Wadhams, P., McLaren, A. S. & Weintraub, R. 1985 Ice thickness distribution in Davis Strait in February from submarine sonar profiles. *J. geophys. Res.* **90**, 1069–1077.

Wadhams, P., Prussen, E., Comiso, J. C., Wells, S., Crane, D. R., Brandon, M., Aldworth, E., Viehoff, T., & Allegrino, R. 1995 The development of the Odden ice tongue in the Greenland Sea during winter 1993 from remote sensing and field observations. *J. geophys. Res.* (In the press.)

Walsh, J. E. & Johnson, C. M. 1979 An analysis of Arctic sea ice fluctuations 1953–1977. *J. phys. Oceanogr.* **9**, 580–591.

Worby, A. P. & Allison, I. 1991 Ocean-atmosphere energy exchange over thin, variable concentration Antarctic pack ice. *Ann. Glaciol.* **15**, 184–190.

Discussion

M. WALLIS (*University of Wales at Cardiff, Cardiff, UK*). Given that the rate of ice formation depends on areas of open water and of thin ice, and is balanced on average by transport through the Fram Strait, would the system not be stable to global warming and constrain possible changes in North Atlantic currents?

P. WADHAMS. There certainly could be a negative thermodynamic feedback along the lines suggested by Dr Wallis, but ice–ocean models show that both the extent and thickness of Arctic sea ice are far more dependent on dynamics than on thermodynamics. Thus it is the effect of global change on Arctic winds that is likely to be the determining factor.

Glaciers in the High Arctic and recent environmental change

By J. A. Dowdeswell

Centre for Glaciology, Institute of Earth Studies, University of Wales, Aberystwyth, Dyfed SY23 3DB, Wales, UK

High Arctic climate change over the last few hundred years includes the relatively cool Little Ice Age (LIA), followed by warming over the last hundred years or so. Meteorological data from the Eurasian High Arctic (Svalbard, Franz Josef Land, Severnaya Zemlya) and Canadian High Arctic islands are scarce before the mid-20th century, but longer records from Svalbard and Greenland show warming from about 1910–1920. Logs of Royal Navy ships in the Canadian Northwest Passage in the 1850s indicate temperatures cooler by 1–2.5 °C during the LIA. Other evidence of recent trends in High Arctic temperatures and precipitation is derived from ice cores, which show cooler temperatures (by 2–3 °C) for several hundred years before 1900, with high interdecadal variability. The proportion of melt layers in ice cores has also risen over the last 70–130 years, indicating warming. There is widespread geological evidence of glacier retreat in the High Arctic since about the turn of the century linked to the end of the LIA. An exception is the rapid advance of some surge-type ice masses. Mass balance measurements on ice caps in Arctic Canada, Svalbard and Severnaya Zemlya since 1950 show either negative or near-zero net balances, suggesting glacier response to recent climate warming. Glacier-climate links are modelled using an energy balance approach to predict glacier response to possible future climate warming, and cooler LIA temperatures. For Spitsbergen glaciers, a negative shift in mass balance of about 0.5 m a^{-1} is predicted for a 1 °C warming. A cooling of about 0.6 °C, or a 23% precipitation increase, would produce an approximately zero net mass balance. A 'greenhouse-induced' warming of 1 °C in the High Arctic is predicted to produce a global sea-level rise of 0.063 mm a^{-1} from ice cap melting.

1. Introduction

Recent general circulation model (GCM) simulations of climatic response to increasing proportions of 'greenhouse gases' in the atmosphere have predicted that Arctic regions will experience enhanced warming relative to lower latitudes, particularly in winter (Cattle & Thomson 1993; Cattle & Crossley, this volume). In addition to these predictions of future, anthropogenically-induced climate change, there is evidence that the Arctic climate has fluctuated over the past few hundred years, presumably in response to 'natural' forcing factors within the climate system (Bradley & Jones 1992). These past fluctuations are exemplified by the warming in many Arctic regions associated with the ending of a period of variable but generally colder conditions, known as the 'Little Ice Age' (LIA), at about the turn of the century (Grove 1988).

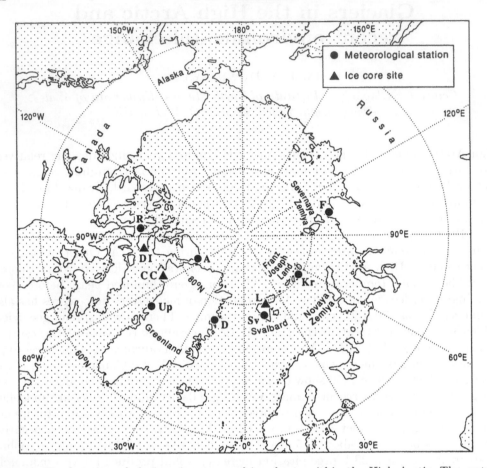

Figure 1. The location of glaciers, ice caps and ice sheets within the High Arctic. The meteorological stations and ice core sites discussed in the text are also shown: (DI) is Devon Island, (CC) Camp Century, (L) Lomonosovfonna, (R) Resolute, (A) Alert, (Up) Upernavik, (D) Danmarkshavn, (Sv) Svalbard Lufthavn, (Kr) Krenk and (F) Fedorova.

The objective of this paper is to examine how ice masses in the High Arctic have been affected by climatic fluctuations in the recent past, and how they may respond to possible future shifts in the Earth's environmental system. Recent climate change is taken here as the past few hundred years, and the next 50–200 years. The latitudinal region forming the High Arctic is defined as that area above 75° N, encompassing the northern Greenland Ice Sheet, the heavily ice covered High Canadian Arctic islands, and the Eurasian Arctic archipelagoes of Svalbard, Russian Franz Josef Land, Severnaya Zemlya and Novaya Zemlya (figure 1). The small De Long archipelago at 76–77° N and 148–158° E completes the circumpolar distribution of ice-covered High Arctic islands. The total area covered by High Arctic ice masses is about 200 000 km^2, or about 37% of the Earth's glaciers and ice caps, excluding the Greenland and Antarctic ice sheets at 1.7 and 12 million km^2, respectively (Meier 1984; Warrick & Oerlemans 1990; Oerlemans 1993).

The paper sets out the observational and proxy records of recent climatic change in the High Arctic using meteorological data and ice cores. Recent fluctuations in Arctic ice mass extent are then examined. Glacier-climate links are investigated through studies involving the measurement and energy balance modelling of glacier mass balance. Finally, the links between High Arctic ice masses and climate are discussed in the context of global sea-level rise.

2. Meteorological records of recent climate change in the High Arctic

(a) Modern meteorological stations

Meteorological data from the High Arctic before World War II are available from only a limited number of stations, reflecting the isolation of this remote, largely ice covered environment. Since the war, observations have been collected from most parts of the High Arctic, although accessible records from the Russian north are fragmentary.

Time series data for several stations around the High Arctic are shown in figure 2. The only available records from the 1930s and earlier are from Upernavik in West Greenland, Isfjord Radio (now relocated to Svalbard Lufthavn) in Spitsbergen, and Fedorova, immediately south of the heavily ice-covered Russian Severnaya Zemlya archipelago (figure 1). The most marked temperature shifts are in the two longest records. Temperatures from 1875 at Upernavik show a warming of 2 °C in mean annual temperature between the last quarter of the 19th century and the 20th, with a particularly rapid rise of about 3.5 °C if the ten years around 1920 are taken alone. The mean annual temperature in Svalbard also rose abruptly by 4–5 °C between 1912 and 1920. However, mean July temperatures at these two stations show a much less marked change over the same periods, of about 20–40% of the mean annual change (figure 2). This is significant for glacier mass balance, because summer temperatures are linked closely to glacier surface ablation, whereas a relative warming between about October and May does not lead to enhanced melting. Even so, these relatively large shifts to warmer conditions are inferred to mark the end of the LIA.

The remaining High Arctic stations in figure 2, with meteorological records of less than 50 years, show high interannual variability, but also some more consistent decadal changes. For example, the interval from 1932 to the mid-1950s was relatively warm at Fedorova, Russia, and the 1960s was a relatively cool period in Svalbard. There was a period of relative summer cooling from about 1964 to 1977 at Resolute, Canada (Bradley & England 1978), followed by warmer temperatures. The period 1920–1950 was particularly warm in West Greenland, followed by cooling of about 1.5 °C in mean annual temperature.

Two concluding points should be made on the basis of the meteorological data in figure 2. First, the records are very noisy in terms of interannual and interdecadal variability. Secondly, the records from some stations show significantly warmer and also cooler intervals during the 20th century, and it is against this background of continuing climatic variability that possible future anthropogenically-induced climate change should be set.

(b) Meteorological records from the mid–19th century

Much of the High Arctic was still being explored and mapped during the 19th century. However, some fragmentary meteorological records do exist from this period, mainly in ships' logbooks. The era of the search for the Canadian Northwest Passage (approximately 74° N), and Sir John Franklin, saw over 25 wintering ships in this area between about 1819 and 1859. Each ship was frozen into the stable winter sea-ice

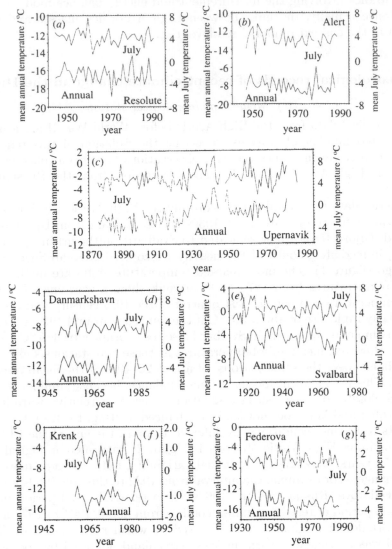

Figure 2. Meteorological records of fluctuations in mean annual and mean July temperature at stations in the Canadian, Greenland and Eurasian High Arctic sectors (located in figure 1). (*a*) Resolute, Cornwallis Island, Canada (74°43′ N, 95°59′ W). (*b*) Alert, Ellesmere Island, Canada (82°30′ N, 62°20′ W). (*c*) Upernavik, West Greenland (72°47′ N, 56°10′ W). (*d*) Danmarkshavn, East Greenland (76°46′ N, 18°40′ W). (*e*) Longyearbyen, Svalbard (78°04′ N, 13°38′ E). (*f*) Krenk Station, Hayes Island on Franz Josef Land, Russia (80°37′ N, 58°03′ E). (*g*) Fedorova, northern Taymyr Peninsula, Russia (77°43′ N, 104°17′ E).

cover, providing a fixed site at sea level for systematic meteorological observations within the LIA (Grove 1988).

The logs of these Royal Navy ships are being analysed (Dowdeswell & Barr 1995). Records of temperature, pressure, wind speed and direction, and sometimes precipitation were logged, with measurements a number of times each day. Instruments were housed away from the ship, elevated from the ground and shaded. Reproducibility has been investigated by comparisons of temperature and pressure data recorded independently by the ships Resolute and Intrepid, locked together in sea ice in 1852–53. Correlation coefficients between the two sets of observations are better than 0.9–0.95.

July temperatures have so far been extracted from the logs of ten ships trapped in shorefast sea ice at locations within 1.5° of latitude 75° N between 1849 and 1859. Mean July maximum temperatures are an average of 4.8±1.4 °C for the ten vessels. July values are presented, since summer temperatures are a first order control on glacier surface melting and on net mass balance. The positions of these ships lie around the modern weather station of Resolute (Bradley & England 1978). Comparing the mean July maximum values for the mid-19th century with modern records shows that only once from 1948–1976 was the observed value below the mid-19th century mean. Modern values are a mean of 7.5±1.3 °C for 1948–63 and 6.0±1.6 °C for 1964–1976. Direct observations from the Canadian High Arctic in the mid-19th century therefore support the evidence of proxy climate datasets, that this was a colder period relative to most of the 20th century, with summers cooler by about 1–2.5 °C.

3. Recent climate change from High Arctic ice-core records

Deep ice cores have been obtained from a number of High Arctic ice caps (figure 3). Where little or no surface melting takes place to complicate core stratigraphy and chemistry, a temporal resolution of ±1–2 years is possible over the past few centuries (Bradley 1985). However, in the Eurasian High Arctic, where mean annual temperatures are significantly higher, melting and refreezing effects dominate and chronology is more difficult to establish (Dowdeswell *et al.* 1990; Tarussov 1992).

Several parameters have been used to investigate climate change in High Arctic ice cores, including: (i) oxygen isotope ratios, which are related to temperature at the time of snow condensation and several complicating factors (Bradley 1985); and (ii) the number and thickness of refrozen melt layers, which are proportional to ice surface melting and therefore provide an index of summer warmth (Koerner 1977). The results of oxygen isotope analyses of ice cores from the High Arctic vary in detail (figure 3), but some broad similarities are present in isotopic records from Devon Island, Canada (Paterson *et al.* 1977), through Camp Century in North Greenland (Johnsen *et al.* 1970), to Lomonosovfonna in Svalbard (Gordiyenko *et al.* 1981) and the Vavilov Ice Dome, Russian Severnaya Zemlya (Kotlyakov *et al.* 1990). The previous two to three centuries have markedly more negative oxygen isotopic ratios than the 20th century (figure 3). The 16th century is somewhat warmer than the 1600–1900 period, but isotopic values do not generally approach those of the relatively warmer 20th century (figure 3). Changes towards less negative ratios over the last 100–150 years are about 1.5 ppt on the Canadian Arctic ice caps, representing a temperature rise of about 2.5 °C over LIA conditions. Core chronologies are most uncertain for the Eurasian High Arctic, due to the intense nature of surface melting.

The recent records of melt layers in Arctic ice cores also show similarities over space, and with the isotopes. On Devon Island the period since about 1860 has shown a rise in the number and thickness of melt layers relative to the preceding 300 years, with summer warming intensifying since the 1920s (Koerner & Paterson 1974; Koerner 1977). Ice cap mass balance is now negative, but is predicted to have been close to zero for the earlier, cooler period (Koerner 1977). The very low frequency of melt layers about 150 years ago indicates that this was the interval of coldest summers

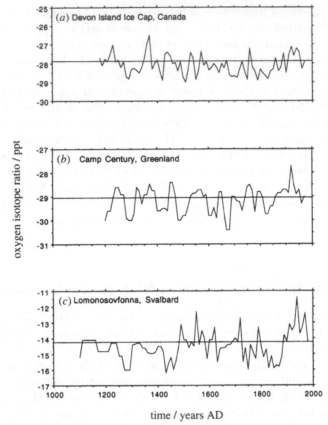

Figure 3. Oxygen isotope ratios over the last thousand years recorded in ice cores from three ice masses in the High Arctic. Horizontal lines in each diagram represent modern isotope values. The core sites are located in figure 1. (a) Devon Island Ice Cap, Canada (from Paterson *et al.* 1977). (b) Camp Century, North Greenland (from Johnsen *et al.* 1970). (c) Lomonosovfonna, Svalbard (from Gordiyenko *et al.* 1983).

over the whole Holocene (Koerner & Fisher 1990). Analysis of melt features in the Dye 3 core, Greenland, suggests a similar cool period from 1530–1860 (Herron *et al.* 1981). In Svalbard ice cores, the interval interpreted to represent from about 1550 to 1920 contains up to 30–40% less refrozen ice layers than that since 1920 (Tarussov 1992). The first half of the 16th century is intermediate between these two periods. The melt layer signal from the large ice caps in the Russian archipelago of Severnaya Zemlya indicates a warming trend from about 120–140 years ago, earlier than in Svalbard (Kotlyakov *et al.* 1989; Tarussov 1992).

The isotopic and stratigraphic evidence from High Arctic ice cores indicates that the cold LIA began to ameliorate from about 1860–1880 in the Canadian Arctic, Greenland and the eastern Eurasian Arctic, but that warming took place somewhat later in Svalbard. Direct meteorological observations from western Svalbard support the ice core evidence on the timing of warming, with temperatures increasing dramatically between 1912 and 1920 (figure 2e). However, despite some broad agreement, the detailed isotopic and stratigraphic records from individual High Arctic ice cores show high interannual to interdecadal variability.

4. Evidence of recent glacier fluctuations in the High Arctic

(a) The retreat of glacier and ice cap margins

Fluctuations in the position of glacier termini can reflect changes in climate, through the effects of shifts in temperature and precipitation on glacier mass balance. However, the links between climate and glacier fluctuations are not simple and glacier dynamic factors must also be considered. Ice masses in most areas of the High Arctic have been retreating for much of the 20th century, whether their margins end on land or in marine waters. On Svalbard, aerial photographs acquired at intervals since the 1930s show that many glaciers are in retreat from clearly defined terminal moraine systems, with the exception of those that have surged. Chronological control for the last few hundred years is usually provided by radiocarbon and lichenometric dating methods, and Werner (1993) has used calibrated lichen growth curves to show that glaciers retreated from prominent moraine systems in north and north-western Spitsbergen from about the turn of the century. In the Canadian High Arctic, a number of glaciers have been retreating over the last century or so. However, the terminus positions of many glaciers and ice caps in Arctic Canada have not changed significantly over the past few decades, although shrinkage and thinning were measured on some especially during the period 1950–1970 (R. M. Koerner, quoted in Meier 1984).

Observational evidence from the 1930s onwards also suggests that ice masses have been retreating throughout most of the Russian High Arctic (37–158° E), and scattered local accounts are available earlier. Comparison of ice margins in Franz Josef Land between aerial photographs obtained in the 1950s and recent satellite imagery suggests a general retreat, in some cases of 2–3 km. In Novaya Zemlya, measurements between the 1930s and 1950s indicated an average retreat of 2–3% (Chizhov & Koryakin 1962). On Severnaya Zemlya, Pioneer Glacier was observed to shrink by 27 km^2 or 11% of its area between the 1930s and 1978, and Govorukha *et al.* (1987) report the loss of about 500 km^2 of ice-covered area throughout the islands between 1931–84. Significant shrinkage of the 80 km^2 ice cover on the de Long archipelago in the Asian High Arctic has also taken place over the last 30–40 years (Verkulich *et al.* 1992). Summarizing these observations, Koryakin (1986) states that during the 20th century Novaya Zemlya has lost 830 km^2 or about 3% of its glacierized area, Franz Josef Land about 700 km^2 or 5%, but suggests, in conflict with Govorukha *et al.* (1987), that ice on Severnaya Zemlya has undergone only minor retreat in limited areas. Together, this amounts to an estimated loss of total ice volume of between 1 200 and 1 600 km^3 this century (Koryakin 1986).

(b) Recent advances of glaciers and ice caps

There are two main exceptions to the broad picture of recent glacier retreat in the High Arctic. Neither is related directly to climate change. First, ice masses in several areas are known to undergo periodic rapid advances or surges, unrelated to climate, punctuating longer periods of stagnation and slow terminus retreat (Meier & Post 1969). Surge-type glaciers are concentrated in some areas, whereas in others the phenomenon does not seem to take place. Within the High Arctic the greatest concentration of known surge-type glaciers is in Svalbard (Dowdeswell *et al.* 1991). The looped moraines indicative of previous surge activity have also been reported from East Greenland (Weidick 1988). Rapid glacier advances resulting from surges are only rarely documented from the Russian and Canadian Arctic archipelagoes (Hattersley-Smith 1969).

Secondly, the marine termini of ice masses are highly sensitive to water depth. This is demonstrated by a strong dependence of the rate of mass loss through iceberg production on water depth (Brown *et al.* 1982). For example, the very rapid retreat of ice through the fjords of Glacier Bay, Alaska, is linked to deep water inland of shallow sills, with glacier margins undergoing very rapid retreat after a small initial change destabilized their position on shallow pinning points (Brown *et al.* 1982). Recent fluctuations of a series of fast-flowing tidewater glaciers in southwest Greenland have been variable. Some of these glacier termini have retreated, whereas others have advanced or remained stable. Warren (1991) relates these variable marginal fluctuations to the influence of differing bathymetry on the dynamic behaviour of individual glacier margins, rather than to climatic variability.

(c) *Glacier response time to climate change*

A time lag between climate change and glacier terminus response is present because mass balance is perturbed throughout glacier length, but is transferred down-glacier at finite velocities over a range of distances. The effect of a climatic shift therefore arrives at a glacier margin over a period, and the terminus position is then a weighted mean of past climate changes over the time interval (T_m) beyond which there is no memory of former climate (Johannesson *et al.* 1989). T_m is the time constant in an exponential, asymptotic approach to a new steady state after a given shift in climate. Johannesson *et al.* (1989) propose that

$$T_m = h/(-b_t), \tag{1}$$

where h approximates to maximum glacier thickness, and b_t is the mass balance at the terminus, which is a negative value. This yields timescales for adjustment to changing mass balance on the order of 10^2 years for many Arctic valley and outlet glaciers, which provides an approximation to the lag time between climatic variations and glacier response. Clearly, the larger ice caps in the High Arctic, where ice may be over 500 m thick and mass loss at the margins is relatively small, will have a longer lag time of a few hundred years.

5. Measurement and modelling of glacier-climate links

(a) *The mass balance of High Arctic glaciers*

The relationship between inputs of mass to High Arctic glaciers, as snow or refrozen meltwater, and mass loss, in the form of meltwater runoff and iceberg production (for tidewater glaciers), is known as glacier mass balance. If, over a balance year, inputs exceed losses, then an ice mass has a positive balance, and *vice versa*. However, long time series of glacier mass balance observations are scarce and losses by iceberg calving are very difficult to quantify, restricting mass balance data largely to ice masses ending on land.

Time series mass balance data are available from several High Arctic ice masses (figure 4). In each region, mean mass balance since the 1950s has been negative, implying a net loss of mass, albeit small in the cases of the Vavilov Ice Dome in Severnaya Zemlya and the Devon Island Ice Cap, Canada. Consistently negative glacier mass balance data have also been reported from Franz Josef Land ice masses

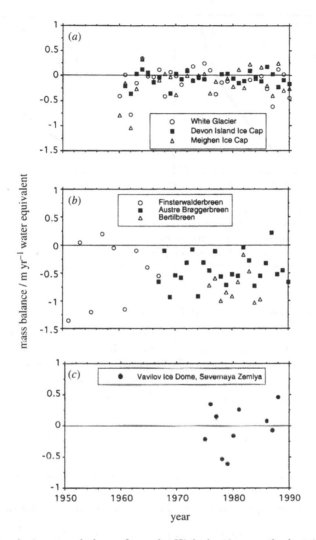

Figure 4. Trends in glacier mass balance from the High Arctic over the last 45 years. The mean and standard deviation of data for each glacier is given in water equivalent. (a) Meighen Ice Cap (-0.13 ± 0.29 m a^{-1}), Devon Island Ice Cap (-0.06 ± 0.13 m a^{-1}), and White Glacier on Axel Heiberg Island (-0.10 ± 0.26 m a^{-1}), Canadian High Arctic (from Koerner (1977) and World Glacier Monitoring Service). (b) Finsterwalderbreen (-0.51 ± 0.59 m a^{-1}), Austre Brøggerbreen (-0.42 ± 0.30 m a^{-1}) and Bertilbreen (-0.74 ± 0.27 m a^{-1}), Svalbard (Hagen & Liestøl 1990). (c) Vavilov Ice Dome, Severnaya Zemlya (-0.03 ± 0.36 m a^{-1}) (Barkov *et al.* 1992).

(Grosswald & Krenke 1962). All the mass balance datasets have a relatively high interannual variability, and it is only in Svalbard that the standard deviations fall entirely below the line representing mass balance equilibrium (figure 4). Even so, it can be concluded that, during the last 30–40 years, glaciers from most areas of the High Arctic have shown no sign of building up; a finding that is in line with meteorological and ice core evidence indicating climate warming over the last century or so (figures 2 and 3).

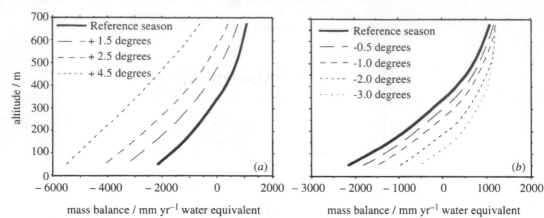

mass balance / mm yr⁻¹ water equivalent

Figure 5. Energy balance modelling of the mass balance of a northwest Spitsbergen glacier in response to recent climate change (after Fleming *et al.* 1995). Modelled mass balance with altitude with: (*a*) increases in annual mean temperature of 1.5, 2.5 and 4.5 °C; (*b*) an envelope of cooler temperatures. Calculations using a modern reference climate are shown for comparison.

(*b*) *Energy balance models of glacier–climate interactions*

Energy balance modelling of the mass balance of High Arctic glaciers provides a method of assessing their sensitivity to climatic shifts. The model, which calculates the components of ice surface energy balance, takes meteorological data, the area distribution with altitude of the ice mass, and parameters defining the global radiation as input values (Oerlemans 1993). The mass balance of a glacier surface is expressed as:

$$M = \int_{\text{year}} [(1 - f) \min(0, -B/L) + P] \, \mathrm{d}t, \tag{2}$$

where M is annual mass balance, f is the fraction of melt water that refreezes instead of running off, B is the energy balance of the surface, L is the latent heat of melting and P is the rate of solid precipitation. The energy balance is found from:

$$B = Q(1 - a) + I_{\text{in}} + I_{\text{out}} + F_{\text{s}} + F_{\text{l}}, \tag{3}$$

where a is the surface albedo, Q is the shortwave radiation reaching the surface, I_{in} and I_{out} are the incoming and outgoing longwave radiations and F_{s} and F_{l} are the sensible and latent heat fluxes. Details of the model are in Oerlemans (1993).

Model results for several Spitsbergen glaciers, using observed meteorological parameters, yielded good predictions of measured mass balance over the last ten years (Fleming *et al.* 1995). Average net balances for 1980–1989, predicted using models tuned to the average for the decade, were −0.44 m and −0.47 m water equivalent for two north-west Spitsbergen glaciers, compared with measured averages of −0.27 m and −0.36 m. The model was then used to predict the effects of recent climate change on glacier mass balance and equilibrium line altitude. Several climate warming scenarios were input to the model (figure 5), which predicted a negative shift in net mass balance of 0.5–0.8 m a⁻¹ for each degree of warming (Fleming 1992), depending on the area/elevation distribution of individual glaciers. By contrast, modelling suggested a cooling of about 0.6 °C, or a precipitation increase of around 23%, would be required to give a zero or slightly positive net mass balance (figure 5).

Table 1. *Predicted contribution of the major High Arctic ice-covered regions to global sea-level rise for a 1 °C warming and adjusted precipitation, according to equation (4). Equation and data on ice cover and precipitation are from Oerlemans (1993)*

region of the high Arctic	ice-covered area / km^2	precipitation / mm a^{-1}	sea-level rise / mm a^{-1}
Svalbard, Arctic Norway	36 600	375	0.018
Franz Josef Land, Russia	13 700	312	0.005
Novaya Zemlya, Russia	23 600	500	0.016
Severnaya Zemlya, Russia	18 300	250	0.005
Ellesmere Island, Canada	80 000	220	0.014
Axel Heiberg Island, Canada	11 700	220	0.002
Devon Island, Canada	16 200	225	0.003
total for the High Arctic	200 100	—	0.063

6. Implications for global sea-level change

The possible contribution of High Arctic ice masses to 'greenhouse induced' global sea-level rise can also be estimated, following Oerlemans (1993). A strong relation has been found between glacier sensitivity and mean annual precipitation, which quantifies the relatively low sensitivity of High Arctic ice masses to climate change compared with glaciers in lower latitudes (Oerlemans & Fortuin 1992). Energy balance modelling experiments on twelve glaciers from varying latitudes and climatic regimes were used to predict mass balance averaged over the entire glacier (B_m). A one degree temperature increase was simulated, in which precipitation was increased in proportion to the saturation vapour pressure calculated from the mean annual air temperature over each glacier (Oerlemans 1993). The experiments gave the following relation concerning mass balance change (∂B_m):

$$\partial B_m = -0.401 - 0.514 \log P, \tag{4}$$

where P is mean annual precipitation, indicating that High Arctic ice masses, with precipitation often of only a few tens of centimetres per year, are less sensitive to climate change than those in warmer and more maritime regions (Oerlemans & Fortuin 1992; Oerlemans 1993).

Applying equation (4) to the ice masses in the High Arctic, excluding the Greenland Ice Sheet, provides an estimate of the contribution of each area to global sea-level rise for a 1 °C warming (table 1). The combined contribution of ice masses in the Canadian, Norwegian and Russian High Arctic gives a global sea-level rise of 0.063 mm a^{-1}. This total represents about 14% of the 0.46 mm a^{-1} sea-level rise predicted by Oerlemans (1993) for the contribution of glaciers and ice caps from all latitudes, excluding Greenland and Antarctica. A problem with these calculations is that values for precipitation are based on few data and, although unlikely to vary by an order of magnitude, will probably require revision as new information becomes available. The complex issue of the influence of the Greenland Ice Sheet on global sea-level is considered elsewhere (Wingham, this volume).

7. Conclusions

Climate change in the High Arctic over the last few hundred years has included the relatively cool LIA, followed by recent warming. However, these broad climatic trends are represented in individual locations by high interannual and interdecadal variability in the climate system (figure 2). Long meteorological records from Svalbard and Greenland show warming of several degrees from the late 19th to early 20th centuries (figure 2). Proxy evidence of High Arctic climate is derived mainly from ice cores, and oxygen isotopic records show generally cooler temperatures (by 2–3 °C) for several hundred years preceding 1900 (figure 3). The proportion of melt layers in ice cores has also risen significantly over the last 70–130 years, indicating recent summer warming. Ice masses in the High Arctic (figure 1), have responded to these changes in climate in a number of ways.

(i) There is widespread geological evidence of glacier retreat in the High Arctic since the end of the LIA. An exception to this is the rapid advance of some ice masses during surges, unrelated to climate forcing.

(ii) Mass balance measurements on High Arctic ice caps since the 1950s show that net balances are either negative or near-zero (figure 4), suggesting that glaciers are responding to recent climate warming.

(iii) Glacier-climate links have been modelled using an energy balance approach to predict the response of glaciers to future and past climate change. For Spitsbergen glaciers (figure 5), a negative shift in mass balance of about 0.5 m a^{-1} is predicted for a 1 °C warming. A cooling of about 0.6 °C, or a 23% precipitation increase, would produce an approximately zero net mass balance.

(iv) A 'greenhouse-induced' warming of 1 °C affecting glaciers throughout the High Arctic is predicted to produce a global sea-level rise of 0.063 mm a^{-1} from ice cap melting (table 1). This is about 14% of the modelled sea-level contribution of all glaciers and ice caps, excluding the Antarctic and Greenland ice sheets.

In conclusion, predictions of the possible future response of High Arctic ice masses to environmental change should be set within the context of continuing climatic variability which is not forced exclusively by anthropogenically-induced changes to the atmospheric system, as exemplified by the direct and proxy records of climate change for the last few hundred years from the High Arctic (figures 2–4).

Parts of this work have been supported by the U.K. Natural Environment Research Council (Grant GR3/8507), the European Community (Grants EV5V-CT93-0299 and INTAS-1010-CT93-0006) and The Royal Society. I thank E. K. Dowdeswell for collating meteorological and mass balance data and R. M. Koerner for comments.

References

Barkov, N. I., Bol'Shiyanov, D. Y., Gvozdik, O. A., Klement'yev, O. L., Makeyev, V. M., Moskalenko, I. G., Potapenko, V. Y. & Yunak, R. I. 1992 New data on the structure and development of the Vavilov Ice Dome, Severnaya Zemlya. *Materialy Glyatsiologicheskikh Issledovaniy* **75**, 35–41.

Bradley, R. S. 1985 *Quaternary Paleoclimatology* Boston: Allen and Unwin.

Bradley, R. S. & England, J. 1978 Recent climatic fluctuations of the Canadian High Arctic and their significance for glaciology. *Arctic alp. Res.* **10**, 715–731.

Bradley, R. S. & Jones, P. D. 1992 Climate since A.D. 1500: Introduction. In *Climate since A.D. 1500* (ed. R. S. Bradley & P. D. Jones), pp. 1–16. London: Routledge.

Brown, C. S., Meier, M. F. & Post, A. 1982 Calving speed of Alaska tidewater glaciers, with application to Columbia Glacier. *US geol. Surv. pro. Paper* **1258** C.

Cattle, H. & Thomson, J. F. 1993 The Arctic response to CO_2-induced warming in a coupled atmosphere-ocean general circulation model. In *Ice in the Climate System* (ed. W. R. Peltier), pp. 579–596. Berlin: Springer-Verlag.

Chizhov, O. P. & Koryakin, V. S. 1962 Recent changes in the regimen of the Novaya Zemlya ice sheet. *Int. Assoc. sci. Hydrol.* **58**, 187–193.

Dowdeswell, J. A. & Barr, W. 1995 Direct observations of 19th century climate in the high Canadian Arctic: the instrumental record from overwintering ships. (In preparation.)

Dowdeswell, J. A., Drewry, D. J. & Simões, J. C. 1990 Comment on: '6000-year climate records in an ice core from the Høghetta ice dome in northern Spitsbergen'. *J. Glaciol.* **36**, 353–356.

Dowdeswell, J. A., Hamilton, G. S. & Hagen, J. O. 1991 The duration of the active phase on surge-type glaciers: contrasts between Svalbard and other regions. *J. Glaciol.* **37**, 388–400.

Fleming, K. M. 1992 Modelling the mass balance of Spitsbergen glaciers. M.Phil. thesis, University of Cambridge.

Fleming, K. M., Dowdeswell, J. A. & Oerlemans, J. 1995 Modelling the mass balance of northwest Spitsbergen glaciers and their response to climate change. (In preparation.)

Gordiyenko, F. G., Kotlyakov, V. M., Punning, Y.-K. M. & Vairmae, R. 1981 Study of a 200 m ice core from the Lomonosov Ice Plateau on Spitsbergen and the paleoclimatic implications. *Polar Geog. Geol.* **5**, 242–251.

Govorukha, L. S., Bol'Shiyanov, D. Y., Zarkhidze, V. S., Pinchuk, L. Y. & Yunak, R. I. 1987 Changes in the glacier cover of Severnaya Zemlya in the twentieth century. *Polar Geog. Geol.* **11**, 300–305.

Grosswald, M. G. & Krenke, A. N. 1962 Recent changes and the mass balance of glaciers on Franz Josef Land. *Int. Assoc. sci. Hydrol.* **58**, 194–200.

Grove, J. M. 1988 *The Little Ice Age.* London: Methuen.

Hagen, J. O. & Liestøl, O. 1990 Long-term glacier mass balance investigations in Svalbard. *Ann. Glaciol.* **14**, 102–106.

Hansen-Bauer, I., Kristensen, M. & Steffensen, E. L. 1990 The climate of Spitsbergen. *Klima Rap.* No. 39/90, pp. 1–40. Oslo: Norsk Meteorologisk Institutt.

Hattersley-Smith, G. 1969 Recent observations on the surging Otto Glacier, Ellesmere Island. *Can. J. Earth Sci.* **6**, 883–889.

Herron, M. M., Herron, S. L. & Langway, C. C. 1981 Climate signal of ice melt features in southern Greenland. *Nature, Lond.* **293**, 389–391.

Johannesson, T., Raymond, C. F. & Waddington, E. 1989 Time-scale for adjustment of glaciers to changes in mass balance. *J. Glaciol.* **35**, 355–369.

Johnsen, S. J., Dansgaard, W., Clausen, H. B. & Langway, C. C. 1970 Climatic oscillations 1200–2000 AD. *Nature, Lond.* **227**, 482–483.

Koerner, R. M. 1977 Devon Island Ice Cap: core stratigraphy and paleoclimate. *Science, Wash.* **196**, 15–18.

Koerner, R. M. & Fisher, D. A. 1990 A record of Holocene summer climate from a Canadian high-Arctic ice core. *Nature, Lond.* **343**, 630–631.

Koerner, R. M. & Paterson, W. S. B. 1974 Analysis of a core through the Meighan Ice Cap, Arctic Canada, and its paleoclimatic implications. *Quatern. Res.* **4**, 253–263.

Koryakin, V. S. 1986 Decrease in glacier cover on the islands of the Eurasian Arctic during the twentieth century. *Polar Geog. Geol.* **10**, 157–165.

Kotlyakov, V. M., Korotkov, I. M., Nikolayev, V. I., Petrov, V. N., Barkov, N. I. & Klement'yev, O. L. 1989 Reconstruction of the Holocene climate from the results of ice-core studies on the Vavilov Dome, Severnaya Zemlya. *Materialy Glyatsiologicheskikh Issledovaniy* **67**, 103–108.

Kotlyakov, V. M., Zagorodnov, V. S. & Nikolayev, V. I. 1990 Drilling on ice caps in the Soviet Arctic and on Svalbard and prospects of ice core treatment. In *Arctic Research: Advances and Prospects* (ed. V. M. Kotlyakov & V. Y. Sokolov), vol. 2, pp. 5–18. Moscow: Nauka.

Meier, M. F. 1984 Contribution of small glaciers to global sea level. *Science, Wash.* **226**, 1418–1421.

Meier, M. F. 1990 Reduced rise in sea level. *Nature, Lond.* **343**, 115–116.

Meier, M. F. & Post, A. S. 1969 What are glacier surges? *Can. J. Earth Sci.* **6**, 807–817.

Oerlemans, J. 1993 Modelling of glacier mass balance. In *Ice in the climate system* (ed. W. R. Peltier), pp. 101–116. Berlin: Springer-Verlag.

Oerlemans, J. & Fortuin, J. P. F. 1992 Sensitivity of glaciers and small ice caps to greenhouse warming. *Science, Wash.* **258**, 115–117.

Paterson, W. S. B., Koerner, R. M., Fisher, D., Johnsen, S. J., Clausen, H. B., Dansgaard, W., Bucher, P. & Oeschger, H. 1977 An oxygen isotope climatic record from the Devon Island Ice Cap, Arctic Canada. *Nature, Lond.* **266**, 508–511.

Tarussov, A. 1992 The Arctic from Svalbard to Severnaya Zemlya: climatic reconstructions from ice cores. In *Climate since AD 1500* (ed. R. S. Bradley & P. D. Jones), pp. 505–516. London: Routledge.

Verkulich, S. R., Krasanov, A. G. & Anisimov, M. A. 1992 The present state of, and trends displayed by, the glaciers of Bennett Island in the past 40 years. *Polar Geog. Geol.* **16**, 51–57.

Warren, C. R. 1991 Terminal environment, topographic control and fluctuations of West Greenland glaciers. *Boreas* **20**, 1–15.

Warrick, R. & Oerlemans, J. 1990 Sea level rise. In *Climate change: the IPCC scienitific assessment* (WMO-UNEP), pp. 257–281. Cambridge University Press.

Weidick, A. 1988 Surging glaciers in Greenland – a status. *Grønlands geol. Undersogelse Rapp.* **140**, 106–110.

Werner, A. 1993 Holocene moraine chronology, Spitsbergen, Svalbard: lichenometric evidence for multiple Neoglacial advances in the Arctic. *The Holocene* **3**, 12–137.

Elevation change of the Greenland Ice Sheet and its measurement with satellite radar altimetry

By D. J. Wingham

Department of Space and Climate Physics, University College London, Holmbury St. Mary, Surrey RH5 6NT, UK

Satellite radar altimetry is presently the only method that has provided the spatial coverage and density of observations needed to reduce the present uncertainty in the mass balance of the Greenland Ice Sheet and its contribution to change in eustatic sea level. The only such measurement reported, however, estimated that southern Greenland was thickening at 23 ± 6 cm a^{-1} which is larger than was thought hitherto. This value is reconsidered given more recent information concerning the errors in the measurement. A survey of measurements of specific mass balance of the Greenland Ice Sheet is given, together with estimates of its sensitivity to temperature change. The expected behaviour is described of errors in the satellite position and errors in the range measurement to the ice sheet surface. The treatment of biases and the number of independent observations of random errors is described. It is found in particular that a higher degree of independence was given to the random errors than should have been the case. The total error is recalculated with this accounted for, and is found to remain dominated by the bias estimate and therefore largely unaffected by this change; the estimate is 23 ± 7 cm a^{-1}. It is concluded that the observation does support a recent thickening of the southern Greenland Ice Sheet.

1. Introduction

The Greenland Ice Sheet is the second largest ice sheet on Earth. It contains 2.7×10^{18} kg of ice, distributed over an area 1.67×10^6 km^2, reaching thicknesses of 3400 m in places (Warrick & Oerlemans 1992). Much of the ice is of Holocene origin, but older, Wisconsin ice with a markedly different rheology is found in deep cores. Each year an estimated 500×10^{12} kg (Ohmura & Reeh 1991) of water is added to the ice sheet by precipitation and an approximately similar amount is lost by surface melting and calving of glaciers into the ocean. This annual exchange with the oceans has a volume equal to a 1.4 mm change in eustatic sea level. Some 200×10^{12} kg of this exchange results from the ablation of ice from the surface of the ice sheet. Should surface temperature increase, the Greenland Ice Sheet is expected to react with a much higher sensitivity than the Antarctic Ice Sheet; what data exist indicate that in rough terms a rise in temperature of 2 K may result in increase in the rate of eustatic sea-level rise equal to 20% of the current exchange (Warrick & Oerlemans 1992).

The uncertainties in these estimates are large because there very few measurements

135

and these relate to relatively small areas of the ice sheet. In these areas, the measurements support a zero net exchange, although the uncertainties are large. The only measurement that surveyed a large section of the sheet was that of Zwally (1989), which indicated the ice sheet in Southern Greenland was thickening by 23 ± 6 cm a^{-1}. The measurement was criticized on both glaciological (van der Veen 1993) and technical grounds (Douglas *et al.* 1990).

In this paper we reconsider Zwally's result in the light of a better understanding now available of satellite orbit reconstruction biases. We describe in §2 measurements of recent changes in the Greenland Ice Sheet and in §3 its sensitivity to climate change. The measurement of Zwally, and possible difficulties with it, is described in §4. In §5 and §6 the behaviour is described of errors in the orbit reconstruction and surface ranging. In §7 we discuss the implication of these behaviours. We find that while some changes to the calculation of the error are required they do not alter the result substantially. We conclude in §8, however, that the result is not inconsistent with what is known and understood about the Greenland Ice Sheet.

2. Present understanding and knowledge of the ice sheet

Ice sheets do not thicken or thin uniformly. Their evolution may be described by the variation with time of the mass per unit area, or *specific mass balance*, which is generally a function of position. This quantity depends on the net mass of ice arriving at a point due to the flow of ice with a depth averaged velocity u, and the rate m at which ice is added or subtracted at the surface, or *specific surface balance*. (This term may possibly include a contribution from the base of the sheet). Taking the density of ice as a constant, (an assumption we return to in §7), conservation of mass provides the following expression for the time evolution of the thickness h of the ice:

$$\frac{\partial h}{\partial t} = -\nabla \cdot (uh) + m, \qquad (1)$$

with m expressed as a height per unit time.

Much of the deformation in an ice sheet occurs at or near the bed, where the shear stresses and temperature are highest. Large changes in the thickness and surface gradient of the ice sheet, which determine the driving stress, or large changes in the temperature at the bed, occur over thousands of years. If one ignores the possibility of rapid changes in the boundary conditions at the bed, one may regard the first term on the right-hand side of (1) as temporally constant, at least as far as changes on decadal time scales are concerned. For our purpose, the consequence of a diverging flow field can be regarded as a constant rate, generally a function of position.

The most complete theoretical calculation of this rate for the Greenland Ice Sheet has been made by Huybrechts (1994). Generally, he finds the present specific mass balance of the ice sheet to be decreasing, as a consequence of the Holocene warming now reaching the more rapidly deforming ice near the base of the sheet. However, he also finds a significant increase in specific balance in South West Greenland, which is not explained simply. The consequences of the replacement of softer Wisconsin ice by harder Holecene ice does not seem as large as a simpler model (Reeh 1985) predicts. Typical magnitudes predicted by Huybrechts (1994) are shown in figure 1.

The specific surface balance of the Greenland Ice Sheet shows strong interannual fluctuations. These fluctuations result principally from year to year fluctuations in

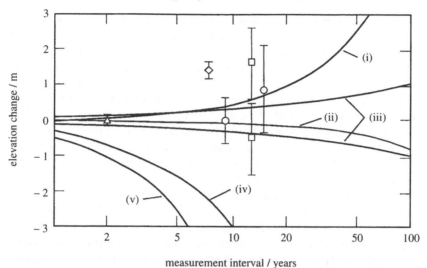

Figure 1. Theoretical predictions and measurements of recent elevation changes of the Greenland Ice Sheet. Theoretical results: (i) long-term trend, south west Greenland (Huybrechts 1994); (ii) long-term trend, Summit (Huybrechts 1994); (iii) growth of the standard deviation in elevation resulting from fluctuations in surface mass balance (van der Veen 1993); (iv) estimate of change in elevation in ablation regions in southern Greenland resulting from an increase in mean summer temperature of 1 K (Braithwaite & Olesen 1990); (v) estimate of average change in elevation at the EGIG equilibrium line, southern Greenland, resulting from an increase in mean summer temperature of 1 K (Ambach 1993). Measurements: ground observations along the OSU and EGIG lines, southern Greenland (Kostecka & Whillans 1988). The measurement interval is equal to the period over which accumulation measurements were made, OSU 9 years, EGIG 15 years; aircraft laser altimeter observations 1991–1993, southern Greenland (Thomas, personal communication); aircraft laser and transceiver observation 1980–1993, OSU line (Thomas, personal communication), satellite radar observation 1978–1985, Greenland south of 72° N (Zwally 1989).

temperature. In zones of accumulation (i.e. regions for which m is positive), these fluctuations are apparent in accumulation measurements from pits and ice cores. In consequence, the specific mass balance will fluctuate randomly with time. Over the time scales of interest here, one may model the fluctuations of m as the sum of a constant rate and a white random process. In consequence, (1) describes an ice thickness that equals the sum of a secular trend and a fluctuation whose variance increases linearly with time. A variance of 10^2 cm^2 is typical of the specific surface balance fluctuations listed by van der Veen (1993) for Greenland. The magnitude of the corresponding standard deviation in thickness calculated from (1) is shown in figure 1.

How do these predictions compare with observation? The experimental measurement of specific mass balance of an ice sheet may be performed by determining separately the terms on the right-hand side of (1). The experiment depends on a theoretical model to determine the depth averaged flow from the measured surface velocities. In fact, very few measurements of this kind have been performed. Typical of these, and regarded as the most careful, are the accumulation zone measurements of Kostecka & Whillans (1988) along transects in central and southern Greenland, termed respectively the EGIG and OSU lines. The line-averaged surface balances that were found are shown on figure 1.

The alternative method is to measure the term on the left-hand side of (1) directly.

This is possible by geodetic means if a change in height may be taken as equal to the change in thickness, which requires that allowance is made for a motion of the bed of the ice sheet relative to the geodetic reference frame. For the timescales we consider, this is a good approximation. Ground based measurements over large distances suffer from uncertainties in datum. In the past two years, however, Krabill and others (unpublished) have made use of a datum established by radio location from satellites. They have repeated aircraft surveys of the EGIG line using a laser altimeter to measure height changes relative to a global ellipsoid. They have also made measurements over the OSU line, where transponder network measurements, an earlier form of satellite positioning system, were available from 13 years previously. Their results are also plotted in figure 1.

Comparing the predicted and measured specific balances, one may conclude that the data do not resolve whether the specific balance is significantly different from zero. However, one must stress, firstly, that there are no observations over much of the ice sheet and, secondly, that the measurements span a comparatively short interval of time. Fig.1 indicates that over a decade of measurement, the specific balance must exceed 10 cm a^{-1} for a secular trend to be observable with statistical significance.

What of the ice sheet as a whole? To obtain the mass balance, one may integrate (2) over the surface of the ice sheet, which leads with the help of the divergence theorem to an expression for the rate of change of volume V of the entire sheet

$$\frac{dV}{dt} = - \int_B d\ell(\boldsymbol{uh}) \cdot \boldsymbol{n} + M. \tag{2}$$

The first term on the right-hand side of (2) is the outward flow at the boundary B, i.e. the calving loss. The second term is the surface balance. According to the calculation of Huybrechts (1994) the present long term trend in mass balance is positive, and expressed as an equivalent specific balance, (i.e. the left-hand side of (2) divided by the area of the sheet), equal to 8 mm a^{-1}. Virtually nothing is known of the spatial correlation of the fluctuations in specific surface balance, and it is possible only to place an upper bound on the fluctuations of M.

The mass balance may be determined experimentally by determining the two terms on the right-hand side of (2). An attempt to do this has been made by Reeh (1985). However, the data are very sparse. From these data Warrick & Oerlemans (1992) place a formal uncertainty of ± 10 cm a^{-1} equivalent specific balance on the ice sheet as a whole.

3. Sensitivity of the ice sheet to climatic change

If the climate changes, the specific surface balance of the ice sheet will change. For the time intervals we consider here, the effect of climatic change on the flow is expected to be negligible. If it is possible to parameterize a changed climate by temperature alone, one may expect the precipitation and ablation to change. The sensitivity of the specific balance to changes in temperature may be estimated from correlation with climatological data. Of course proceeding this way can take no account of larger changes in atmospheric circulation that may occur in a changed climate.

Of the two effects, the change in ablation with temperature has had the greater attention.

Braithwaite & Olesen (1990) have deduced a value of -49 cm K^{-1} a^{-1} and

-62 cm K^{-1} a^{-1} for the change in specific balance from data collected at two sites on the margins of the ice sheet in south west Greenland. Ambach (1993) has perturbed an equation expressing the surface heat balance using differential coefficients determined from observations made in 1959 along the EGIG line. He calculated changes in specific surface balance of -33 cm K^{-1} a^{-1} at the equilibrium line, (i.e. the locus of points at which m equals zero), falling to -83 cm K^{-1} a^{-1} at lower altitudes. The resulting changes in elevation with time in ablation zones predicted by Braithwaite & Olesen (1990) and Ambach (1992) are shown in figure 1.

Turning to the ice sheet as a whole, there are too few observations to make a secure estimate of the sensitivity of its mass balance to temperature. Both Ambach (1993) and Braithwaite & Olesen (1990) have estimated the sensitivity of eustatic sea-level to the surface temperature of the ice sheet by simply assuming their specific sensitivities extend over its whole area. However, there is no basis for determining the errors arising from this assumption. Warrick & Oerlemans (1992), quoting the result of a study by Oerlemans, give 7 ± 5 cm K^{-1} a^{-1} for the equivalent specific mass balance sensitivity with an allowance for an increase in precipitation. These authors do admit the similarity between their estimates and others is due at least in part to the dependence on the same observations.

While the errors in these estimates are poorly constrained, their order of magnitude does indicate that the mass balance of the Greenland Ice Sheet is considerably more sensitive to temperature changes than is the mass balance of the far larger Antarctic Ice Sheet. This arises from the fact that surface ablation, which is strongly dependent on temperature, accounts for 40% of the mass turnover in Greenland. In Antarctica surface ablation is usually regarded as negligible because surface temperatures are very much lower.

4. Measurement of the mass balance by satellite

There is then uncertainty as to the contribution of Greenland to the present change in eustatic sea level. There is also uncertainty as to the consequences of climate change. Direct observation constrains the present contribution to ± 0.4 mm a^{-1}; a 2 K rise in temperature may lead to a rise in contribution of 0.6 ± 0.4 mm a^{-1}. The uncertainties will change significantly only with observations of considerably greater spatial extent and density.

In 1989, Zwally (1989) gave the results of observations of the average change in elevation of the Greenland Ice Sheet south of $72°$ N. The observations were made by radar altimetry from two satellite platforms, Seasat and Geosat, over the 7 year interval 1978–1985. The elevation change given by Zwally was 23 ± 6 cm a^{-1}, corrected for isostatic adjustment of the bed. This value was calculated from 5906 observations widely distributed over southern Greenland and was therefore a direct measurement of the mass balance of the southern section of the ice sheet.

Zwally's value is plotted in figure 1 for comparison with estimates of specific balance. The value is larger than any observed specific balance, and is larger than may be explained by random fluctuations in specific surface balance, even assuming these are perfectly correlated over the entire ice sheet. The measurement has been criticized on these grounds by van der Veen (1993).

The mismatch between Zwally's result and those of ground based observations needs closer investigation. Satellite altimetry is presently the only method that provides the spatial extent and density of measurements that are needed to reduce the uncertainty in present knowledge. The question that arises is: are the estimates of error attached to the satellite measurements too small, and, if so, over what time interval need measurements be made to arrive at an improved and statistically significant estimate of the mass balance of the Greenland Ice Sheet? We shall deal with the first of these in §5 and §6, and the second in §7.

A radar altimeter determines the elevation of the ice surface relative to an ellipsoid by taking the difference of two measurements, the distance between a known point on the satellite and the ellipsoid, and the distance between the same point on the satellite and the nearest point on the surface of the ice sheet. See, for example, McGoohan (1975) for a description of the principles of operation. The observations are made along the ground track of the satellite. As the satellite continues in its orbit, a network of ground track⠄ are laid down. In this way, the ice sheet as a whole may be sampled. In principle, a change can be measured by simply repeating the measurement. In practice, the network of tracks do not generally repeat exactly. Observations of changes are made at locations where tracks from each network cross each other, known as *cross-over points* or simply cross-overs.

The error ε in the elevation difference at the cross-over may be written

$$\varepsilon = \varepsilon_0 + \varepsilon_r, \tag{3}$$

where ε_0 arises from errors in the distance to the closest point on the ice sheet. One needs to consider, for each of these errors, whether they may be regarded as having an expectation of zero? If this is not the case, then systematic errors may lead to the false inference of a secular trend. One needs also to consider the degree to which they are uncorrelated, for this determines the reduction in the random error when the elevation differences at the cross-overs are averaged to determine the ice-sheet-wide change. The character of the errors ε_0 and ε_r are quite different, and we shall consider them separately.

5. Errors in the reconstruction of the orbit

The position of a radar altimeter satellite is determined by combining a model of the forces experienced by the satellite with observations of the position and speed of the satellite at various times around the orbit, in order to calculate the position at other times. The error ε_0 arises from this procedure because (i) the force equations are non-linear and errors arise in their integration even when the initial conditions are known perfectly, (ii) the initial conditions, provided by the observations, are known only imprecisely, and (iii) the forces, principally those due to gravity and drag, are known only imprecisely. An element of empiricism enters the calculation. The orbit is computed over a time interval, typically several days, with the drag terms chosen so as to minimize the difference between the calculated and observed satellite positions. The resulting modelled orbit is termed an *arc*.

Consider first the proposition that the error ε_0 has a zero mean. It is known (Engelis 1988) that elevation differences at cross-overs contain a stationary term related to the satellite orbit and errors in the gravity force model. In Zwally's initial work, the orbits of the two satellites were computed with different force models, which would increase this term, but this difficulty was later removed (Zwally 1990). The error is a slowly varying function over the Earth with a value that may exceed 1m, a stronger longitudinal dependence than latitudinal, and a gradient that may be estimated to rarely exceed 2 cm per degree of longitude (see, for example, Jolly & Moore 1994). On the basis that no correction was made for this error, Douglas *et al.* (1990) criticized the result of Zwally. This criticism is often quoted. In fact, Zwally subtracted from the elevation difference a value of 0.4 m, equal to the apparent height difference in sea-level in the region around Greenland measured by Seasat in 1978 and Geosat in 1975.

We consider now the degree of correlation between the error at different cross-over points. The time-variant component of the orbit error is dominated by a harmonic variation with a period equal to the time taken by the satellite to complete one orbit (Engelis 1988). The amplitude and phase of this variation change slowly relative to this period. It is usually assumed that they are uncorrelated arc-to-arc. Within an arc, they may be highly correlated (Chelton & Schlax 1993). An error of this character in the orbit projected onto the satellite's ground tracks may be expected to vary only slowly with latitude and longitude. My own investigation of the spatial correlation function of these errors, using observations from the ERS-1 and TOPEX/Poseidon altimeters, is not complete at the time of writing. Nonetheless, this study indicates that for a region the size of Greenland, the assumption that the in-arc time-variant errors are strongly spatially correlated is a better one than the assumption that they are uncorrelated. On this basis, the number of independent observations of the time-variant orbit error in the cross-over set is closer to the number of arcs than the number of cross-overs.

6. Errors in the range to the ice sheet

The distance between the satellite and the nearest point on the ice sheet is measured by timing the delay experienced by an electromagnetic pulse travelling from the satellite and returning to the satellite as an echo from the surface. An error arises in this measurement because of (i) unknown variations in the speed c of the pulse through the atmosphere, (ii) noise on the echo, and (iii) non-uniqueness in the relationship between the shape of the radar echo and the distance to the nearest point on the ice sheet.

The speed of the pulse depends on the surface pressure, the electron content of the ionosphere, and the integrated water content of the atmosphere. The dependencies on the first two of these are accounted for with numerical models, the latter from observations of the atmosphere. Over time intervals of days the assumed dependencies are in error. However, over the three-month interval of the two sets of satellite observations, the error in an averaged elevation change from these sources is a few centimeters at most. See, for example, Wagner & Cheney (1992) for a closer discussion. The noise in the measurement results principally from the noise-like character of radar echoes. This source of error is not significant to large averages of the kind we consider here.

To describe the third source of error needs some explanation of the range measurement. The transmitted pulse has a duration Δ. The echo is observed as a function of time. At any instant t the echo power is proportional to the product of the area of the surface that lies a distance between $\frac{1}{2}ct$ and $\frac{1}{2}c(t + \Delta)$ from the satellite, and a function that depends on the antenna of the radar and its attitude with respect to the Earth. (We ignore here effects on the echo shape due to small scale fluctuations in the surface termed roughness). If the shape of the surface is known, the evolution

D. J. Wingham

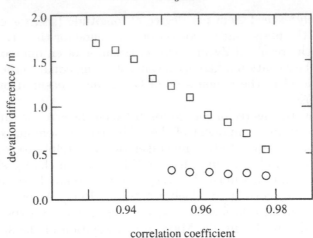

correlation coefficient

Figure 2. The standard deviation of elevation differences observed by the ERS-1 satellite altimeter over 35 days in 1992 at crossing points of the ground track of the satellite orbit, plotted as a function of the mean value of the maximum of the cross-correlation function between the four echoes measured closest to the crossing point. 8081 cross-over locations in East Antarctica; 1878 cross-over locations in the central tropical Pacific. Correlations less than 0.95 between echoes from the ocean are very improbable. The standard deviation of the ocean cross-overs is mainly the result of time-variant orbit reconstruction error. The cross-over elevation differences determined by retracking ice sheet echoes contain a source of error additional to those from the ocean. The error increases as the shape of the echoes near the crossing point become more dissimilar as a result of the irregular topography of the ice sheet.

of this area with time is known. The distance to the nearest surface location may then be determined from the echo delay. If the shape of the surface is known only approximately, however, the problem arises as to how to determine the distance to the nearest point, that is a number, from a function that is only approximately known. This is the situation one faces with ice sheet altimetry, (see, for example, Robin *et al.* 1983). This difficulty has been dealt with by the introduction of a rule, termed (confusingly) a retracker, that associates to the echo a delay, from which a range is determined. The relationship between the range so determined and the range to the nearest point on the surface remains obscure; indeed, closer inspection shows it to be non-unique. Nonetheless, provided the shape of the surface does not change during the measurement interval, and the radar remains unchanged, the cross-over elevation difference should determine a pure displacement of the surface in the direction of the satellite.

The error associated with the shape of the echo arises in the following way. Echoes are observed at discrete points along the satellite ground track. These points do not generally coincide with the location of an orbit cross-over, but lie within a few hundred metres of it. The echo shapes at each point, and hence the delays assigned by the 'retracker', may differ solely as the result of the irregular shape of the ice sheet. The cross-over elevation difference then contains an error that is due to the irregular topography of the ice sheet. This is clearly demonstrated in figure 2. The figure demonstrates that, along with the error due to the time variant component of the orbit, ice sheet topography is a dominant source of error in the cross-over differences.

Is it reasonable to assume the error associated with the topography has a zero mean? One may answer this question by treating the topography as a random variable. Provided that the topography at the times of the two measurements differed solely by a displacement in the direction of the satellite, and that the radar instrument did not change, one concludes very simply that the elevation difference is unbiased. To deal with the question of correlation, one may argue that the error in each cross-over difference is uncorrelated, since the topography giving rise to this error is correlated on scales of ten kilometres, less than the distance separating the cross-over locations for the most part.

One needs to consider the two assumptions of these arguments. The topography on scales of ten kilometres is associated with the flow of ice over irregularities in its bed. This will not change over the time interval we consider. The second assumption raises potentially more serious difficulties. The problem of dealing with the topographic error when different radars are used has only recently been solved (Wingham 1994). The Seasat and Geosat radars were very similar. The Geosat radar, however, was mounted on a satellite whose attitude was neither sufficiently stabilized nor measured. The problem of determining theoretically the error in a ice sheet cross-over difference due to an uncorrected attitude is not tractable in a usefully general way. One can say, from a practical standpoint, that the consequence of an attitude error is usually to increase the range by a few cm or so. Thus, if this Geosat-related error is present in Zwally's measurement, one would expect on these grounds that it would reduce the average cross-over distance.

Another feature of the radar interaction that needs comment is the penetration of the surface by the radar wave (Ridley & Partington 1988). The near surface layers of the Greenland Ice Sheet, particularly in the south, change as a result of thermodynamic processes. It is at least possible that these changes may result in the false deduction of a height change from the echo shapes. The problem is a difficult one to analyse in detail because there is recent theoretical evidence (Winebrenner, personal communication) that the altimeter radar wavelength is such as to make the echo sensitive to grain size and layering. Against this, it is an empirical observation that the echo shapes from Greenland are very stable season to season. It must also be borne in mind that the altimeter echo is an average over length scales very much larger than those of the processes effecting change in the surface layers. My own experience is that effects from this source rarely exceed 10 cm and have a variance dominated by seasonal time scales.

7. Discussion

Zwally (1989) gives the root-mean-square (RMS) of the elevation differences at the Seasat–Geosat cross-overs as 126 cm, the Geosat–Geosat cross-overs as 147 cm and the Seasat–Seasat cross-overs as 100 cm. Seasat ocean cross-over differences at that time had an RMS of 80 cm. Assuming this to be the time-variant orbit error, one concludes that the topographic contribution to the ice sheet cross-over difference error was 60 cm. With this value, one can calculate the RMS Seasat orbit error at 56 cm and the RMS error of the Geosat orbit at 94 cm.

In §5 we noted that Zwally corrected for a simple bias between the Seasat and Geosat orbit solutions. The given uncertainty in this bias was 40 cm. The question remains as to whether a constant is an adequate representation for the functional form of the error. Since Greenland extends over 20° of latitude an argument can be made, on the basis of our description of this error in §5, that a bias as large as 20 cm could remain after the subtraction. In §5 we also argued there were fewer independent observations of the time-variant orbit error than the 5906 observations. The Geosat arcs were of five days duration, and the cross-overs were a set formed from observations over three months, which is 18 arcs. Certainly some decorrelation occurs within each arc; we shall take the number of independent observations of the time variant orbit error as 100. One then calculates the time variant orbit error contribution to the Seasat–Geosat cross-overs as 14.5 cm. In §6 we argued that the number of independent observations of the topographic error was equal to the number of observations. The topographic error is then reduced in the average to 0.8 cm. Finally, regarding these four sources of error as independent, one obtains a final uncertainty of 46 cm, compared with Zwally's 41 cm.

The total remains dominated by the error in the bias correction. The fact that Zwally (1990) obtained a result that differed by 20 cm with a different gravitational force model supports the value we have attributed to this source of error.

As shown in figure 1, Zwally's result is particular because it is the only result that indicates that large parts of the ice sheet are not in balance. However, given that (i) the error bars in figure 1, and the envelope of random fluctuations, are drawn at 1 standard deviation, (ii) the ground observations are very limited in their coverage, and (iii) there is theoretical evidence of differential changes over the ice sheet, it is our view that the altimeter measurement is not contradicted by other observations.

The assumption in §2 that the near-surface density is a constant function of depth and time is not met in practice. Newly fallen snow has a significantly lower density that glacier ice. If the increase in height observed by altimetry was due to a contemporary increase in snowfall, this would result in a larger change in height than given by (1). However, van der Veen (1993) found that records from central Greenland could only explain at most an increase of a few centimetres.

There is another line of argument that requires comment. The present measurements of the Earth's polar wander are sometimes used to argue that the Greenland Ice Sheet cannot be thickening at this rate. First, we observe that the satellite measurement provided no coverage of half of Greenland. Huybrecht's predictions, which we discussed in §2, indicate differential behaviour over the ice sheet. Second, we observe that the rate of polar wander is affected by a number of different processes that are redistributing mass around the Earth, and to describe the present wander one has a number of degrees of freedom, including the distribution of ice at the last glacial maximum and the present distribution of sources and sinks of water. Peltier (1988), for example, accounted for a newly recognized source of water, melting from small ice caps and glaciers, by adjusting the distribution of ice at the last glacial maximum. What is needed for this line of argument to be convincing is a study that determines the uniqueness with which any one of these sources may be constrained.

8. Conclusions

We have reconsidered the result of Zwally (1989) that the southern part of the Greenland Ice Sheet is thickening at 23 ± 6 cm a^{-1}. While we have questioned his treatment of random errors, we have not been able to find an argument that would greatly change the measurement. To hold otherwise requires an argument for explaining why the experiment generated differential results over the ocean and ice sheets in excess of 1 m, and we are unable to do this. The measurement supports the proposition that the southern part of the Greenland Ice Sheet thickened from 1978–1975 at a higher rate than one can account for by random fluctuations in surface mass balance. Van der Veen has considered that the increase is due to an increase of accumulation over the past few centuries. He has concluded that the available evidence does not support this possibility. If this is correct, the observed increase in thickeneing is due to the mass flow field. This conclusion is supported in sign by the theoretical prediction of Huybrecht (1994), although the measured magnitude is larger.

What is needed is more observations. Since Zwally's result the ERS-1 satellite has provided coverage of Greenland to 82° N from 1991–1994. The ERS-2 satellite will be launched in early 1995 with coverage planned to 1998. Following these, a third European satellite will be launched in 1998. The United States/French satellite TOPEX/Poseidon has been in flight from 1992. It has demonstrated orbit reconstructions with an accuracy of 3 cm. The satellite provides coverage up to 68° N and the observations of this satellite are likely to be of smaller direct importance for determining the mass-balance of the Greenland Ice Sheet; nonetheless the satellite may be of great value in reducing the errors in the orbit reconstruction of other altimeter satellites. This sequence of satellites will provide a decade of coverage. Combined with Seasat observations, a measurement spanning two decades should be possible by 1998. In short, there will be many opportunities test to increase the number of observations of elevation change of the Greenland Ice Sheet.

I would like extend my grateful thanks to Bob Thomas, of the NASA, Washington, DC, who kindly gave me details of the laser altimeter observations of Bill Krabill, of NASA Wallops Flight Centre, and to Duncan Curtis of my own laboratory, who provided me the data shown in figure 2 in advance of its submission for publication.

References

Ambach, W. 1993 Effects of climatic perturbations on the equilibrium-line altitude, West Greenland. *J. Glaciol.* **39**, 5–9.

Braithwaite, R. J. & Olesen, O. B. 1990 Increased ablation at the margin of the Greenland Ice Sheet under a Greenhouse-effect climate. *Ann. Glaciol.* **14**, 20–22.

Chelton, D. B. & Schlax, M. G. 1993 Spectral characteristics of time-dependent orbit errors in altimeter height measurements. *J. geophys. Res.* **98**, 12579–12600.

Douglas, B. C., Cheney, R. E., Miller, L., Agreen, R. W., Carter, W. E. & Robertson, D. S. 1989 Greenland Ice Sheet: is it growing or shrinking? *Science, Wash.* **248**, 288.

Engelis, T. 1988 On the simultaneous improvement of a satellite orbit and determination of sea surface topography using satellite altimetry. *Manuscr. Geod.* **13** , 180–190.

Gundestrup, N. S., Bindschadler, R. A., and Zwally, H. J. 1986 Seasat range measurements verified on a 3-D ice sheet. *Ann. Glaciol.* **8**, 69–72.

Huybrechts, P. The present evolution of the Greenland Ice Sheet: an assessment by modelling. *Global planet. Change* **9**, 39–51.

Jolly, G. W. & Moore, P. 1994 Validation of empirical orbit error corrections using cross-over difference differences. *J. geophys. Res.* **99**, 5237–5248.

Kostecka, J. M. & Whillans, I. M. 1988 Mass balance along two transects of the west side of the Greenland Ice Sheet *J. Glaciol.* **34**, 31–39.

McGoogan, J. T. 1975 Satellite altimetry applications. *IEEE Trans. Microwave Theory Tech.* **MTT-23**, 970–978.

Ohmura, A. & Reeh, N. 1991 New precipitation and accumulation maps for Greenland. *J. Glaciol.* **37**, 140–148.

Peltier, W. R. 1988 Global sea level and Earth rotation. *Science, Wash.* **240**, 895–901.

Ridley, J. K. & Partington, K. C. 1988 A model of satellite radar altimeter return from ice-sheets. *Int. J. rem. Sens.* **9**, 601–624.

Reeh, N 1995 Was the Greenland Ice Sheet thinner in the late Wisconsinan than now? *Nature, Lond.* **317**, 797–799.

Robin, G. de Q., Drewry, D. J. & Squire, V. A. 1983 Satellite observations of polar ice fields. *Phil. Trans. R. Soc. Lond.* A **309**, 447.

van der Veen, C. J. 1993 Interpretation of short-term ice-sheet elevation changes inferred from satellite altimetry. *Clim. Change* **23**, 383–405.

Warrick, R. & Oerlemans, J. 1992 Sea level rise. In *Climate change: the IPCC scientific assessment* (ed. J. T. Houghton, G. J. Jenkins & J. J. Ephraums), pp. 257–281. Cambridge University Press.

Wingham, D. J. 1994 A method for determining the average height of a large topographic ice sheet from observations of the echo received by a satellite altimeter. *J. Glaciol.* **41**, 125–141.

Zwally, H. J. 1989 Growth of Greenland Ice Sheet: measurement. *Science, Wash.* **246**, 1587–1589.

Zwally, H. J. 1990 Reply to: Greenland Ice Sheet: is it growing or shrinking? *Science, Wash.* **248**, 288.

Permafrost and climate change: geotechnical implications

By Peter J. Williams

Geotechnical Science Laboratories, Carleton University,
1125 Colonel By Drive, Ottawa, Canada K1S 5B6

The behaviour of the ground in the cold regions of the world is characterized by freezing and thawing. The porous and particulate nature of soils presents conditions for phase change which lead to their unique properties and behaviour in cold climates. Accordingly, the periodic and unstable nature of atmospheric climate and of surface microclimate produces characteristic disturbances in the near-surface layers of the ground in the cold regions. These include the consequences of melting such as subsidence (thermokarst topography) and instability of slopes (landslides, mudflows), as well as the thermodynamic and mechanical effects of freezing, especially frost heave. Frozen soils show temperature-dependent creep (some forms of solifluction and deformation of foundations) and continuing heave (expansion of ground over long periods of time). These effects have important geotechnical implications for the design of highways, airports, buildings and, notably, pipelines. The complexity of the design problems for major structures, especially pipelines, has not been widely understood.

If there is global warming due to anthropogenic emissions of gases this will influence the direction and intensity of the ground disturbances, the nature of which has been recognized over the last three or four decades. However the effects of such warming due to atmospheric climate change will only become apparent over many decades. In the short term they will be masked by other ground temperature changes due to microclimatic effects and to inter-annual variability of climate and weather. Over a period of a century or more, if warming trends continue, there will be important modifications of terrain and physiography.

1. Introduction

Transitions between the solid and liquid phases of water are responsible for most of the properties and behaviour of earth materials which are particular to the cold regions (Williams & Smith 1991). This is of fundamental importance in understanding the consequences of thermal disturbance such as climate change. The transitions do not occur only at the normal freezing point but to several degrees below that temperature. Particle surface forces and capillarity cause the freezing point of the water in the soil to fall as ice forms. The water remaining has a thermodynamic potential such that water moves along temperature gradients both within the frozen soil and in adjacent unfrozen material. Ice accumulates as a result, in the form of segregations in the frozen soil.

These segregations are excess ice, so-called because, when thawed, the water will be in excess of that which the void space of the soil can accomodate. The formation of the excess ice gives frost heave. This causes stresses and deformations which disrupt foundations for constructions of all types (see e.g. Johnston 1981, Phukan 1985). Thawing of the ground containing excess ice leads to break-up of roads, slope failures and disturbance by subsidence, of foundations for buildings, pipelines and other constructions. Current research is largely geotechnical in application (Burn & Smith 1993).

Corresponding effects occur under natural conditions giving rise to various forms of slope instability and the development of numerous features of surface relief including small hillocks and mounds and various forms of so-called patterned ground (Washburn 1978).

The creep properties of frozen soil are less well-known than the strength loss associated with thaw but are important particularly where complex structures may be harmed by deformation of frozen ground. The creep properties relate to the ice in the frozen soil. The nature of the bonds between ice and particles and between ice and the unfrozen, adsorbed water are also important (Ershov 1996). As the amounts of unfrozen water change in response to temperature changes or applied stresses, gradients of potential cause migrations of water and associated translocation of ice. The creep properties are highly dependent on temperature, especially in the range from freezing point to $-2\,°C$ or $-3\,°C$.

2. Relation of ground temperatures to atmospheric climate

The implications of climate change in the Arctic for constructed foundations of all types, centre on temperatures within the ground and their importance in relation to the characteristic behaviour described above.

Temperatures in the near-surface ground depend broadly on the atmospheric climate but are modified by the complex processes of energy exchange that operate continuously between atmosphere and subsurface. Thus the strictly local surface conditions: vegetation or other cover, topographic form and aspect, and the thermal properties of the soils themselves (Williams & Smith 1991) are responsible for the ubiquitous local variations in ground temperature and, notably, in the mean annual ground temperature.

Once the temperatures for the soil surface have been established, the precise temperature values with depth and time depend essentially on the thermal diffusivity of the earth materials in question. The seasonal (annual) temperature wave moves down into the ground but it weakens with depth, is delayed and reaches only to 15 to 20 m depth or much less if there is a latent heat exchange due to freezing and thawing. Longer-term variations in the atmospheric climate, over decades, centuries, millenia and longer, are transmitted to greater depths, the depth usually being roughly proportional to the square root of time from onset. Permafrost (figure 1) is the condition where ground remains frozen year in year out. It is not, of course, 'permanently' frozen – indeed it is the formation (aggradation) and disappearance (degradation) of permafrost that are geotechnically the most important consequences of temperature change. In addition to the excess ice formed by moisture migration, large bodies of excess ice of various origins occur frequently in permafrost. Mean annual ground temperatures (a value less than $0\,°C$ defines the presence of permafrost), commonly differ from mean annual air temperatures by several degrees but the amount varies. Where the ground temperatures generally are around $0\,°C$ permafrost occurs in a discontinuous or patchy fashion.

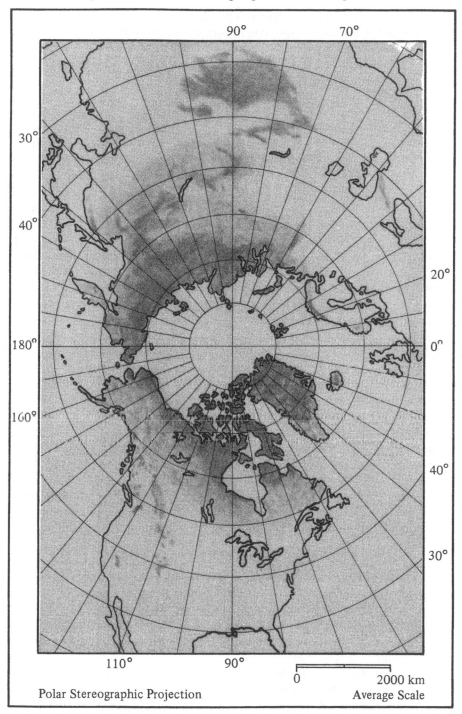

Figure 1. Permafrost in the Northern Hemisphere. The intensity of shading indicates the degree of continuity, the lightest indicating occasional permafrost which underlies only limited, discrete areas in the terrain (from Williams & Smith 1991).

Summer thawing extends to a depth depending on the mean annual temperature and amplitude of the annual temperature wave in the ground surface. In the absence of permafrost, winter freezing, which extends deeper under colder conditions but rarely to more than a metre or two, is an important geotechnical consideration.

3. Role of microclimatic disturbance

Modifications of the ground surface and its cover, the causes of which may be natural or anthropogenic, disturb the energy balance and lead to compensating changes in ground temperatures. The effects of damaging or removing vegetation, for example, in the permafrost regions, are usually to allow the arrival of more heat to the surface and deeper summer thawing. If there is as a result, sufficient subsidence (consolidation) that standing water occurs, then the thaw process is accelerated still further by the heat-absorbing properties of the water and the increased thermal conductivity of the more saturated materials. When the mean annual temperature of the ground surface rises above 0 °C the permafrost starts to thaw from the surface down and given time, all of the permafrost will disappear. Destruction of the natural vegetation by passage of vehicles is perhaps the classic example of environmental damage in the Arctic. The subsidence and standing water that mark sometimes a single vehicle pass continue to enlarge.

Changes of atmospheric climate inevitably lead to changes in microclimate. For example, heavier snow fall will give a deeper, insulating snow cover and thus higher winter (and thus mean annual) ground temperatures. In areas where the permafrost is discontinuous and relatively warm (within a degree or two of 0 °C) it can be absent where winter snow cover is deeper than elsewhere. Further examples are climate-induced changes to the vegetation cover, modification to wind patterns or to soil moisture conditions, any of which affect the energy balance and thus the temperature of the ground. Conceivably a change of atmospheric climate might lead only to a compensating change of ground temperature (to re-establish the energy balance at the ground surface) but so complex is the ground surface energy exchange, modification of other components as well is to be expected. Koster *et al.* (1994) and Brown (1994) review the extensive literature on climate change in relation to permafrost.

4. Evidence for ground warming

The effects of warming of the ground are easily recognized in the cold regions. The consequences of human activity disturbing the microclimate serve as analogues for natural effects which modify ground thermal conditions and illustrate the nature, extent and rapidity of terrain disturbance. In the absence of surface disturbance by human activity, the origins of obvious warming, whether microclimatic or climatic, are often unclear.

If there are large bodies of ice in the permafrost their melting leads to subsidence and if this is widespread the terrain is referred to as thermokarst. Such wet, irregular and (during its formation at least) unstable terrain covers tens of thousands of square kilometres particularly in the Arctic regions of the former Soviet Union and (although to a lesser extent) Arctic and sub-Arctic Canada. It is reasonable to assume much of this warming was due to a changing atmospheric climate, rather than to local, microclimatic effects.

There are also areas of thermokarst too large to ascribe to purely local microclimatic effects yet which are not so widespread or uniform to attribute them with certainty to global atmospheric effects. They may be the result of climate changes which are less than global in extent and microclimatic effects can also be extensive, as for example, the long-term effects of destruction of vegetation by forest fires (an increased incidence of forest fires may itself reflect climate change).

Mean annual air temperatures have commonly shown increases of 1–2 °C in this century. Temperatures in the ground reflect such changes of atmospheric climate and Lachenbruch *et al.* (1988), for example, describe temperature profiles for Alaska which indicate a 2–4 °C warming at the permafrost surface in this century. Some profiles however indicated that cooling had occurred during the past decade. Meteorological observations on the other hand show that cooling occurred between 1940 and 1960 over much of the State (Osterkamp & Lachenbruch 1990). The somewhat confusing details are similar to what is reported for other locations: attempts to correlate climate history with ground temperature profiles have had only limited success (particularly as depth increases). Furthermore, the irregular distribution of inferred temperature changes suggests that whatever the connection may be with anthropogenic atmospheric pollution, it is not a simple one. Changes in precipitation or other components of the soil moisture regime have various complicating effects considered further below (§ 6).

General circulation models predict larger changes of atmospheric climate in the future although there is considerable uncertainty; Karlén *et al.* (1993) for example, stress the inherent variability of climate through historical time. Relatively conservative prognoses are that global temperatures will increase by 2 °C over the next 100 years with possibly several times that figure for the polar regions (IPPC 1990). If such occurs, similar temperature increases can be expected to follow generally, although not necessarily uniformly, within the ground. There will also continue to be, simultaneously, occurrences of warming of the ground and thaw, due to strictly microclimatic effects, and in fact, microclimatic effects will continue to produce cooling and new freezing in particular situations.

5. Geotechnical consequences of environmental change

In the cold regions, the geotechnical properties and behaviour of earth materials are extremely sensitive to changes in the environment, whether natural or anthropogenic, in a manner without counterpart elsewhere on the Earth's surface. The cold regions extend further than the polar regions, with extensive permafrost at much lower latitudes in interior continental, high altitude situations, e.g. northern China (figure 1). As a whole, the geotechnical problems are amongst the economically most significant aspects of cold climates. In the absence of freezing, temperature changes of earth materials usually have little effect on their geotechnical properties. But at freezing temperatures, especially those close to 0 °C, temperature effects become the dominant design consideration. The effects of freezing or thawing have long been reasonably successfully overcome for most structures by relatively simple but costly engineering procedures (Johnston 1981; Phukan 1985). The worst situations in the terrain will if possible be avoided by relocation of the structure. Replacement of soils at risk from frost heave is widely used, for example in highway and airport construction. A design procedure of using permafrost as a foundation material involves insulation to prevent thawing and the use of piles extending below the depth of any expected thaw. In the last two decades however, new demands have arisen, most notably (both geotechnically and in terms of cost) in engineering for oil and gas extraction and particularly for pipelines (Williams 1989).

Where permafrost is absent, then foundations for buildings, highways, airfields etc. need to be designed to avoid the effects of seasonal freezing only. This extends to a depth of only a metre or two (rarely, somewhat more) and the large ice inclusions found in permafrost cannot occur. A general warming of the ground will reduce the depth and thus is of little concern. A cooling may increase the depth and threaten water mains and other conduits, or the stability of road beds. Such structures are however, often built with a significant factor of safety recognising the uncertainties surrounding the depth of annual freezing.

Permafrost provides a (relatively) stable base for foundations. If however, the mean ground temperature is only just below 0 °C, a correspondingly small increase of ambient temperature will cause thawing-out, which advances downwards and, after some time, very slowly from the base of the permafrost upwards as a result of the geothermal heat flux. Subsidence and loss of bearing capacity cause extensive damage to structures but ultimately the ground should be more stable after the permafrost has retreated.

The permafrost does not have to disappear to provide serious problems. Even in the high arctic, where the mean ground temperature may be −10 °C or even colder, a degree or two rise of the mean annual ground surface temperature will cause an increase in summer thawing of 10 cm or so while in less cold situations the annually thawing layer may increase substantially, to a metre or more. This will disrupt foundations to the extent they are dependent on the bearing capacity of the thawed layer – for example in roads or airstrips. The underlying permafrost hinders drainage weakening the soil.

Recently numerical simulations (Riseborough & Smith 1993) indicate that where the permafrost layer is only some tens of metres thick (that is, its temperature is near 0 °C) there can be large fluctuations (a metre or more) in the level of the top of the permafrost over say, a ten year period. The effect is a consequence of the year by year variability of climate (weather) coupled with the greater conductivity of frozen ground compared to unfrozen. It occurs independently of climatic trends and is an element of the inherent problems of geotechnical engineering in permafrost regions.

If there is a continuing warming, atmospheric in origin, the permafrost will disappear from large areas. During the process the zone of the most unstable, that is, the warmest permafrost will move northwards, bringing difficult geotechnical conditions to areas where, previously, the permafrost was a relatively more reliable foundation material. The possible extent of such changes is considered below (§6 and §7).

The warming of frozen ground (as opposed to its thawing) by reducing its creep resistance, reduces the bearing capacity of piles or other footings in permafrost (Nixon 1990). Thus failures may occur even though the pile is still within permafrost, if the temperature rises to −1 or −2 °C.

A perhaps unexpected result of warming is renewed heaving of the frozen soil at such temperatures because of the higher content of unfrozen water and greater permeability. Indeed, Ershov (1996), Cheng (1983) and Mackay (1988) all describe heaving and ice accumulation occurring while thawing of the frozen ground (proceeding from the surface down) is taking place. Water migrates towards the lower temperatures to freeze there ('continuing heave', §6), even though the system is undergoing warming. This mobility of water in the frozen soil as it warms is also important in respect to movements of pollutants which are often assumed in geotechnical practice to be contained by frozen soils.

6. Geotechnical problems for the oil and gas industry

The economic significance of pipelines, their high cost and the special geotechnical challenges such extended structures pose, warrant particular consideration. The most sophisticated geotechnical designs yet required in cold regions engineering are those for pipelines.

Oil pipelines, which are inherently warm, are liable to cause thawing and subsidence. Gas pipelines, on the other hand, may be designed to carry gas chilled to below 0 °C in permafrost regions (to avoid thawing the ground) and are liable to produce further freezing of the soil. As pipelines are normally only a metre or so below the surface, most of the effects outlined above are very important.

The TransAlaska oil pipeline was successfully constructed with the pipe above the ground, supported on steel piles which often contain a self-cooling device (cryosiphon), to avoid thawing ice-rich permafrost. The piles reach to depths of some 6–20 m. The belated recognition of the need for such constructions resulted, however, in a rise in costs of billions of dollars (Williams 1989).

The major difficulty for gas pipelines occurs where the permafrost is discontinuous (meaning that it is absent locally because of microclimatic conditions) and the pipe carrying the chilled gas initiates freezing. At such transitions between previously frozen and newly freezing soil and also whereever there is a change in the type of soil, differential frost heaving occurs. Large-scale experimental studies (Williams *et al.* 1993) with a section of 27 cm diameter pipe modelling these circumstances showed stresses and deformations that indicate the operation of a pipeline could be hampered within a few months of construction. The same experiments showed that heaving (accumulation of ice) occurs slowly in a more than 50 cm thick layer of already frozen soil (under the low temperature gradients prevailing around the pipe, and the near 0 °C temperatures). This example of 'continuing heave' which could be predicted from earlier studies (Burt & Williams 1976; Miller 1980; Ohrai & Yamamoto 1985 and others), illustrates the source of the high stresses which develop in the course of months and represent the continuing and longer term threat to gas pipelines in frozen ground.

If the ground warms, apart from the displacements that result from complete thawing, both the changes in strength and creep properties of frozen ground and the renewed heaving associated with temperatures rising to near 0 °C can be important. The problems described above arising from differential heave at transitions between frozen and unfrozen soil and between types of soil, will probably be worse if there is a changing externally imposed thermal regime.

Any change causing a modification to the water balance such that soil moisture levels increase, a modification of microclimate, of weather or of atmospheric climate may also lead to increased frost heave. The gradient along which the water moves to the freezing layer where the cryosuctions develop, depends on the water pressure conditions in the underlying layers. Heaving will be greater when the water table is elevated. Even though the extent of soil frozen may be less, the concentrations of ice will be greater and, depending on the structure, are more likely to result in dangerous deformations and stresses.

No major gas pipeline has yet been built in the North American Arctic (essentially because of cost) and Russian experience has not been good. Pipes have been variously overstressed, deformed, lifted from the ground and exposed by erosion following freezing and thawing (figure 2).

Figure 2. Large pipeline exposed by erosion following freezing, heaving and thawing of near surface layers above permafrost (Siberia). Photograph by E. D. Ershov.

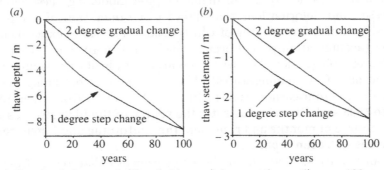

Figure 3. (a) Depth of thaw and (b) subsidence ('thaw settlement') over 100 years, resulting from a step change of temperature of 1 °C and from a steady temperature rise of 0.2 °C per decade, calculated for soil with 30% excess ice.

7. Relative importance of microclimatic and climatic effects

There are many microclimatic effects, both natural and anthropogenic, which demonstrably cause more abrupt and in the short term greater thaw and subsidence (known as thaw settlement) of the ground than those arising from atmospheric climate trends. This is illustrated by figure 3 which shows the amount of subsidence as a function of time, for two cases: one involving a microclimatically induced thawing, the other thawing due to progressive atmospheric climate change. The microclimatic change is taken to be a step change, for example loss of vegetation cover, which results in a warming of the ground surface, essentially complete within a year or two, by 1 °C (the value might well be greater). A simple approach based on Stefan's equation (Lunardini 1981) was used for the calculations of depth of thaw with the assumption that the permafrost had a temperature of just 0 °C. Although Stefan's equation is not very satisfactory for calculations for specific field locations, it illustrates well the principles involved. Over a century very significant subsidence will have occurred in both cases for ice-rich ground (30% excess ice was assumed) but initially it is much greater for the microclimatic, step change condition. The subsidence becomes equal at 100 years in this example, as a consequence of the assumed values which lead to equal total cooling effects (the thawing index, the product of time and positive temperatures, °C a), for the two cases at that time.

Pipelines usually have an expected operational life of some decades. Over a period of thirty to forty years the effects of warming due to atmospheric climate change are likely to be less than and not easily separable from those due to interannual variability and to microclimate as discussed in previous sections (including those disturbances associated with the construction of the pipeline itself). Thus one can conclude that serious though the soil thermal problems are for pipelines, they are not primarily the result of atmospheric climate change.

Planning for pipelines for fifty or a hundred years hence, however, must take into account the conditions that may prevail at the time. The evolution of the terrain should be an important factor in the timing of pipeline developments for extraction from particular fields. For example, the ice-rich, and relatively warm Yamal area of Russia, already regarded as one of the most difficult, could be considerably more unstable and flooded a century from now and partly below sea level.

8. Long term and extreme scenarios

Because very extensive areas of permafrost have temperatures near to 0 °C, a change of climate which leads to a rise of mean air temperatures of, say, 2 °C, will ultimately cause the disappearance of permafrost from hundreds of thousands of square kilometres of the earth's surface. In mainland Canada (Heginbottom 1994) over 25% of the area underlain by permafrost has permafrost with temperatures warmer than −2 °C. Such permafrost predominates in a broad belt (the 'sub-Arctic') of some 1000 km, south to north, and occurs, decreasingly and more locally, up through the Northwest Territories to the mainland coast. There would be a widespread northward extension in disturbances of the terrain (subsidence, landslides etc.) during the warming wherever the permafrost contained ice in excess of the pore space. Similarly, Nelson & Anisimov (1990) believe that thermokarst formation will be very extensive in Western Siberia and little permafrost left there if the mean air temperature rises 5 °C.

Only with a rather extreme rise in temperature, of perhaps 12 °C, would all the Eurasian permafrost thaw and then only after hundreds of millenia for the deepest parts. But more important is the fact that some 25% of the area has mean ground temperatures of −5 °C or higher (Ershov 1996) and thus would thaw under a temperature rise of that amount. Even small temperature increases of one or two degrees over a century, would eventually have widespread effects and under the more extreme warming scenarios major physiographic changes would occur in that time.

Large quantities of water are held in the permafrost as ice. Although a planned world ice inventory was not completed, it is estimated (Shumskiy & Vtyurin 1963 in UNESCO 1967)) that 0.2–0.5×10^6 km^3 or about 1% of the total world volume is held in frozen ground (99% being in glaciers and icecaps). In some regions, 50 to 80% of the upper 20–30 m of the ground is ice. Of this perhaps a third is excess ice which would be expelled as water with corresponding subsidence of the ground when thaw is complete. Vtyrin (1978) concludes that over the greater part of the permafrost area of the former Soviet Union the excess ice content averages out to 1–2 m of thickness. In the far North, notably the Yamal (an important gas extraction region) and northern Sakha regions the figure is 5–10 m. Much of this ice is fairly near the ground surface and would therefore thaw within a century or two (compare figure 3) if surface temperatures increase by a few degrees. A corresponding amount of subsidence would occur and, in many respects, the effects would be similar to those with a rise in sea level, with extensive flooding, coastal erosion etc. Under the more extreme scenarios significant rises in sea level are, of course, also envisaged. The released water should give rise to an increase in river discharge and an increase in landslides, erosion generally and flooding, together with increased sediment loads in the rivers.

Widespread thawing of permafrost may lead to increased release of methane. The complex problems of release of methane, carbon dioxide or other gases under changing near-surface soil conditions are not well understood (Christensen 1991). The break-down of clathrates has geotechnical implications. These are hydrates of methane which occur at depths of hundreds of metres, in the polar regions, under combinations of high pressure and low temperature (Judge *et al.* 1994). They are both a hazard in drilling operations and a potential source for natural gas. There would generally be a long interval of time (thousands or tens of thousands of years) before a warming at the surface would be transmitted sufficiently to the depths where the clathrates occur, increasing the possibility of a transfer into the gas phase. However there is gas being released currently where marine transgression produced a sudden temperature rise hundreds or thousands of years ago.

Were temperatures to ultimately stabilize at a few degrees higher than present there would be significant positive consequences in those regions lying closest to but outside the (reduced) permafrost regions. These regions would have smaller costs for housing, transportation and infrastructure generally. Also within the coldest parts there would be some benefits. Many marine operations for example, would be favoured by a longer ice-free season (Gerwick 1990).

9. Conclusions

In the Arctic and the cold regions more generally, the thermal-mechanical regime, the stability, of the ground is sensitive to anthropogenic disturbance of two distinct types: direct effects due to disturbance of the natural ground surface and therefore of the surface energy exchange, and less direct effects from changes of atmospheric climate. But analogous processes are occurring naturally and extensively and it can be difficult to distinguish natural from anthropogenic effects.

Ground warming following from anthropogenic atmospheric climate change, as presently estimated, is relatively slow compared to warming due to microclimatic disturbances or in association with interannual variability of climate. Thus for geotechnical constructions such as pipelines (and indeed the majority of structures in the cold regions) the lifetime of which is a few decades, climatic warming does not currently represent a major problem additional to the recognized and geotechnically very demanding, problems of cold climates in general.

Relating to this is an important conclusion from a scientific point of view. Precisely because the near-surface layers in the permafrost regions are so susceptible to temperature change from various causes, they do not, contrary to a more generally held view, represent a sensitive indicator over the short term, of atmospheric climatic change.

Ultimately, if temperatures continue to rise, over the next century, by the predicted values of 2 °C or more, there will be a migration northwards of the areas most affected by terrain instability, through hundreds of thousands of square kilometres. The subsidence, landslides, and loss of bearing capacity will be similar to those occurring at present in regions with extensive warm permafrost and which is the basis of current scientific research for the geotechnical and environmental problems of the cold regions.

Refinements of the global models for predicting future climates are, from the geotechnical viewpoint, less pressing than improving understanding of the ground-atmosphere interface and of the ground itself. Geotechnical problems arising from microclimatic instability are with us now and always will be. If the more extreme climatic scenarios come about, mitigative procedures would have to be applied on a larger scale and in elaborate and extensive construction works to combat effects due to thawing over the next century. Because of the extent and geographical location of ground that will be involved, additional problems of flooding, coastal erosion and river sedimentation etc. will occur. But there will also be ultimately an improvement in geotechnical conditions in areas in which permafrost is no longer present.

I am grateful for the hospitality of the Scott Polar Research Institute, for discussions with colleagues there and the extensive resources of its unique library in the preparation of this paper. D. W. Riseborough and J. A. Heginbottom have kindly discussed at length their recent studies of permafrost temperatures and distribution, and the former carried out the calculations for figure 3.

References

Burn, C. R. & Smith, M. W. 1993 Issues in Canadian Permafrost Research. *Progress phys. Geog.* **17**, 150–172.

Burt, T. P. & Williams, P. J. 1976 Hydraulic conductivity in frozen soils. *Earth Surf. Processes* **1**, 349–60.

Brown, J. (ed.) 1994 Permafrost and climate change: The IPA report to the IPCC. *Frozen Ground* **15**, 16–26.

Cheng, G. 1983 The mechanism of repeated segregation for the formation of thick layered ground ice. *Cold Reg. Sci. Tech.* **8**, 57–66.

Christensen, T. 1991 Arctic and sub-arctic soil emissions: possible implication for global climate change. *Polar Record* **27**, 205–210.

Ershov, E. 1996 *General Geocryology. Studies in Polar Research.* Cambridge University Press. (In the press.) (Translation of *Obschee Geokryologia*.)

Gerwick, B. C. 1990 Effect of global warming on Arctic coastal and offshore engineering. *J. Cold Reg. Engng* **4**, 1–5.

Heginbottom, J. A., DuBreuil &, M. A., Marker, P. A. 1994/5 Map: Canada – Permafrost, scale 1:7500,00. In *National Atlas of Canada* (5th edn). Canada: Department of Natural Resources. (In the Press.)

Intergovernmental Panel on Climatic Change Report, quoted in: *Canadian Climate Digest 1991: Climate change and Canadian impacts: the scientific perspective*, CCD 91-01. Canada: Atmospheric Environment Service, Environment Canada.

Johnston, G. H. (ed.) 1981 *Engineering design and construction.* New York: John Wiley and Sons.

Judge A. S., Smith, S. L. & Majorowicz, J. 1994 The current distribution and thermal stability of natural gas hydrates in the Canadian polar regions. In *Proc. 4th. Offshore polar Engng Conf., Osaka*, vol. 1, pp. 307–314.

Karlén, W., Friis-Christensen, E. & Danielson, B. 1993 *The Earth's climate – natural variations and human influence*. Stockholm: Elforsk AB.

Koster, E. A., Nieuwenhuijzen, M. E. & Judge, A. S. 1994 Permafrost and climate change: an annotated bibliograpy. *Glaciological Data Report* **27**, WDC-A for Glaciology. Boulder, CO: University of Colorado.

Lachenbruch, E., Cladouhos, T. T. & Saltus, R. W. 1988 Permafrost in Alaska and predicted global warming. *Proc. 5th Int. Permafrost Conf., Trondheim*, vol. 3, pp. 9–17.

Lunardini, Virgil J. 1981 *Heat Transfer in Cold Climates*. New York: van Nostrand Reinholdt.

Mackay, J. R. 1983 Downward water movement into frozen ground, Western Arctic Coast, Canada. *Can. J. Earth Sci.* **20**, 120–134.

Miller, R. D. 1980 Freezing phenomena in soils. In *Applications of Soil Physics* (ed. D. Hillel). Academic Press.

Nixon, J.F. 1990 Effect of climatic warming on pile creep in permafrost. *J. Cold Reg. Engng* **4**, 67–73.

Ohrai, T. & Yamamoto, H. 1985 Growth and migration of ice lenses in partially frozen soil. *Proc. 4th. Int. Symp. Ground Freezing, Sapporo, Japan*, vol 1, pp. 79–84. Rotterdam: Balkema.

Osterkamp, T. & Lachenbruch, A. 1990 Thermal regime of permafrost in Alaska and predicted global warming. *J. Cold Reg. Engng* **4**, 38–42.

Phukan, A. 1985 *Frozen ground engineering*. Englewood Cliffs, NJ: Prentice-Hall.

Riseborough, D. W. & Smith, M. W. 1993 Modelling permafrost response to climate change and climate variability. In *Proc. 4th Int. Symp. thermal Engng Sci. Cold Regions. USACRREL, Hanover, NH*, pp. 179–187.

UNESCO 1970 *Perennial ice and snow masses*. UNESCO/IASH.

Vialov, S. S. 1965 *Rheological properties and bearing capacity of frozen soils*, translation 74 (from the Russian). Hanover, NH: United States Army Cold Regions Research and Engineering Laboratory.

Vtyurin, B. I. 1973 Patterns of distribution and a quantitative estimate of the ground ice in the USSR. *Int. Conf. Permafrost, USSR Contrib.*, 159–164.

Washburn, A. L. 1978 *Geocryology*. London: Edward Arnold.

Williams, Peter J. & Smith, M. W. 1991 *The frozen Earth. Fundamentals of geocryology*. Cambridge: Cambridge University Press.

Williams, Peter J. 1989 *Pipelines and permafrost: science in a cold climate* (2nd edn). Ottawa: Carleton University Press.

Williams, P. J., Riseborough, D. W. & Smith, M. W. 1993 The France-Canada joint study of deformation of a pipeline by differential frost heave. *Int. J. offshore pol. Engng* **3**, 56–60.

Discussion

M. WALLIS (*University of Wales at Cardiff, Cardiff, UK*). The evaporation from water-logged land decreases air temperatures by around 1–2 °C, comparable to the 2 °C rise considered by Prof Williams. This decrease depends on winds, humidity and boundary layer mixing. So permafrost thawing over 100 km-sized regions would cause more than microclimate changes – e.g. cloud cover and runoff. Emissions of methane are limited via its oxidation by methanotrophizing the water surface layer, and at temperatures close to zero, air-to-water transfer of oxygen is more effective relative to biogeneration: i.e. permafrost swamps will probably generate much less methane than do tropical swamps.

P. J. WILLIAMS. Feedbacks of the kind mentioned by Dr Wallis are complex, numerous, progressive and important especially over decades or centuries. My point, of course, is that geotechnical activity produces microclimatic disturbances which are gross and sudden, if localised, and such effects are crucial in the design of pipelines, for example, even if these are to be relatively short-lived.

Greenland ice core records and rapid climate change

By J. A. Dowdeswell[1] and J. W. C. White[2]

[1] *Centre for Glaciology, Institute of Earth Studies, University of Wales, Aberystwyth, Dyfed SY23 3DB, Wales, UK*
[2] *Institute of Arctic and Alpine Research and Department of Geological Sciences, University of Colorado, Boulder, Colorado 80309, USA*

Long ice cores from Greenland yield records of annually resolved climate change for the past ten to twenty thousand years, and decadal resolution for one hundred thousand years or more. These cores are ideally suited to determine the rapidity with which major climate changes occur. The termination of the Younger Dryas, which marks the end of the last glacial period, appears to have occurred in less than a human lifetime in terms of oxygen isotopic evidence (a proxy for temperature), in less than a generation (20 years) for dust content and deuterium excess (proxies for winds and sea-surface conditions), and in only a few years for the accumulation rate of snow. Similarly rapid changes have been observed for stadial-interstadial climate shifts (Dansgaard–Oeschger cycles) which punctuate the climate of the last glacial period. These changes appear to be too rapid to be attributed to external orbital forcings, and may result from internal instabilities in the Earth's atmosphere–ocean system or periodic massive iceberg discharges associated with ice sheet instability (Heinrich events). In contrast, the Holocene climate of the Arctic appears to have been relatively stable. However, the potential for unstable interglacials, with very rapid, shortlived climatic deteriorations, has been raised by results from the lower part of the GRIP ice core. These results have not been confirmed by other ice cores, notably the nearby GISP2 core. Evidence from other records of climate during the Eemian interglacial have yielded mixed results, and the potential for rapid climate change during interglacial periods remains one of the most intriguing gaps in our understanding of the nature of major Quaternary climate change.

---

1. Introduction

Ice cores taken from the crests of large ice masses provide a continuous and high resolution record of climate change. Time series of up to several hundred thousand years, on a number of environmentally related parameters, have been obtained from ice cores over 3 km long drilled recently through the Greenland and Antarctic ice sheets (GRIP project members 1993; Grootes *et al.* 1993; Jouzel *et al.* 1993). Each successive annual increment of snowfall is buried and preserved in the cores, together with any contaminants, and variations in parameter values can be used to infer the nature of past environmental changes; for example, the oxygen isotopic composition of the ice yields information on palaeo-temperatures, air-bubbles provide evidence on the changing composition of the atmosphere, acidity levels record past volcanic eruptions, and dust levels provide an index of storminess (Johnsen *et al.* 1995). It is the combination of long record length and high time resolution that makes ice cores one of the most effective sources of evidence on both the nature and the rate of environmental change on timescales of up to 10^5 years.

Two deep ice cores, the Greenland ice-core project (GRIP) and Greenland Ice Sheet project 2 (GISP2) cores, have recently been drilled by European and American teams at or close to the crest of the Greenland Ice Sheet. In this paper, the environmental record provided by these two Arctic ice cores is summarized, with particular emphasis on the very abrupt way in which climatic parameters have switched between relative warmth and cold on several timescales over the last 100–150 000 years. The rapidity and magnitude of these climatic shifts is in marked contrast to the relatively more subdued climatic changes of the past 10 000 years, and has served to emphasize the likely existence of significant thresholds in the behaviour of the linked atmosphere-ocean-cryosphere system. The relationship between these changes in climate at the summit of the Greenland Ice Sheet and the time-dependent behaviour of the North Atlantic ocean circulation is also discussed, together with possible correlations with climate change in North-West Europe.

2. The GRIP and GISP2 ice core records

(a) *Core sites and chronology*

The GRIP and GISP2 ice cores were drilled at or close to Summit, in central Greenland, in the early 1990s. The surface elevation at these sites is about 3200 m and the ice is over 3000 m thick. The GRIP core site is at the modern ice divide, whereas the GISP2 site is 28 km to the west. Optimal ice core sites are placed exactly at the ice crest, because here there is no horizontal component to ice flow, and thus ice deformation, especially close to the bed, does not take place as long as the ice divide itself does not migrate through time.

The timescales for the two ice cores are derived from the counting of annual layers of snow accumulation for the upper part of each core (e.g. back to about 14 500 years ago for GRIP). Deeper in the cores a modified steady-state ice flow model is used to calculate the age-depth relationship (i.e. the rate of core thinning resulting from depth-dependent vertical strain), assuming that the stratigraphy is undisturbed. Again taking the GRIP core as an example, model calculations are pinned to two characteristic features observed in other dated climate records (figure 1); first, the termination of the Younger Dryas event at 11 500 years ago and, secondly, marine isotope stage 5d at 110 000 years ago (Dansgaard *et al.* 1993).

(b) *General description of ice core stratigraphy*

Oxygen isotopic values and electrical conductivity measurements (ECM) for the whole length of the GRIP and GISP2 ice cores have been presented and compared in some detail (Dansgaard *et al.* 1993; Grootes *et al.* 1993; Taylor *et al.* 1993). The oxygen isotope profiles are expressed as $\partial^{18}O$ values, representing the relative difference in parts per thousand between the $^{18}O/^{16}O$ abundance ratios of ice core samples and standard mean ocean water (SMOW). The $\partial^{18}O$ value of snow is taken to reflect its temperature of formation, with higher $\partial^{18}O$ values indicating higher temperatures (Johnsen *et al.* 1989). ECM data vary with ice acidity, which can reflect volcanic

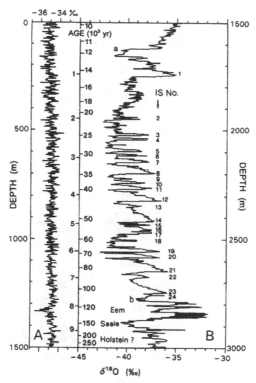

Figure 1. The GRIP $\partial^{18}O$ record plotted on a linear depth scale (from Dansgaard *et al.* 1993 and Johnsen *et al.* 1995). A represents the Holocene record, established by counting annual ice layers. B is the record from 10 000 years to more than 250 000 years, with a chronology derived from ice flow modelling. IS are interstadials.

events and changes in the concentration of alkaline dust (Taylor *et al.* 1993). Colder periods tend to be more dusty, leading to ice of lower conductivity.

The $\partial^{18}O$ and ECM records from the two Greenland drill sites are consistent for the upper 90% of the cores, but vary markedly in the 10% closest to the ice sheet bed (Grootes *et al.* 1993; Taylor *et al.* 1993). The core records can be divided into three parts. First, the upper 1500 m or so of the cores represents the Holocene or last 10 000 years. It is an interval of relatively stable climate, with mean $\partial^{18}O$ values of -34.7 ppt and -34.9 ppt for the GISP2 and GRIP cores, respectively (Grootes *et al.* 1993). The Holocene warm period is separated from the underlying isotopically more negative ice, indicating colder conditions, by abrupt shifts in both $\partial^{18}O$ and ECM values (figure 1). Detailed correlation between the two cores within the Holocene is difficult due to the small magnitude of Holocene isotopic fluctuations (less than 2 ppt). When compared with the remainder of the cores, the Holocene is seen as a period of unusual climatic stability (figure 1).

The second part of the cores is that between about 1500 m and 2750 m in depth, or about 103 000 years ago according to the GRIP timescale (Johnsen *et al.* 1995). Oxygen isotopic and ECM evidence suggests very much more variable climatic conditions for this part of both cores, where inter-core correlation is good down to 2750 m. This interval represents the last glaciation, with a series of colder stadials and warmer interstadials, similar to marine oxygen isotope stages 2 to 5d. Each of the 24 interstadials (IS) identified in the GRIP core is labelled in Figure 1. Only the most recent

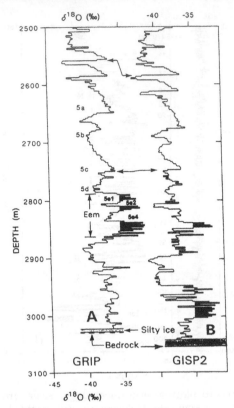

Figure 2. Comparison of the GRIP (A) and GISP2 (B) $\partial^{18}O$ records from the deepest parts of the ice cores (from Johnsen *et al.* 1995). $\partial^{18}O$ values higher than today are shaded black. Substages within marine isotope stage 5 are indicated. Note that ice of high $\partial^{18}O$ values is much closer to bedrock in the GRIP 2 than in the GRIP core (Johnsen *et al.* 1995).

interstadial, beginning at about 14 500 years ago, approaches the isotopic values of the Holocene, and this is rapidly replaced by the isotopically more negative, cold conditions of the Younger Dryas (figure 1).

The remainder of both cores, below about 2750 m or so, shows a period over which oxygen isotope values return to a level equal to and sometimes higher than those of the Holocene. This is interpreted as the last, or Eemian interglacial, occurring at between about 115–130 000 years ago. Below this, and closest to the base of the cores, is ice more negative in $\partial^{18}O$. This is assumed to represent older ice from the Saalian glaciation and beyond.

Three key points can be made about the Greenland ice core records of the Eemian interglacial. First, and arguably the most novel inference from the $\partial^{18}O$ record in the GRIP ice core, was that of continuing climatic instability within the Eemian (figure 1). Unlike the Holocene, the last interglacial contained rapid isotopic shifts of up to 5 per mille, and the colder parts of the Eemian were isotopically similar to interstadials within the last glacial. This represents temperatures about 5 °C cooler than the Holocene (GRIP project members 1993). Secondly, the three warmest intervals of the Eemian were up to 4–5 °C warmer than those of today (Johnsen *et al.* 1995). These two contrasts between the Holocene and the Eemian appeared to have occurred despite the fact that the isotopic (e.g. deuterium excess) and chemical (e.g. concentrations of sodium, magnesium, sulphate and nitrate) signatures of the warm stages of the Eemian are similar to those of the Holocene, suggesting a generally similar atmospheric circulation.

A third point, however, has cast some doubt on the interpretation of the Eemian record in the GRIP ice core. When the $\partial^{18}O$ and ECM records from the GRIP and GISP2 cores were compared, there was a lack of detailed correlation below about 2750 m (figure 2), although some general trends are common to both cores (Grootes *et al.* 1993; Taylor *et al.* 1993). This implies that ice flow deformation is likely to have affected the chronology and stratigraphy of the lowest 10% of one or both of the cores in order to account for these observed differences. Inclined ice layering, observed below 2678 m and 2847 m in the GISP2 and GRIP cores, respectively, may be an indicator of deformed ice (Grootes *et al.* 1993; Taylor *et al.* 1993). The modern location of the GRIP core at the exact crest of the Greenland Ice Sheet means it is less susceptible to folding at depth than the GISP2 site, but modelling suggests that during the last glacial period the ice divide may have migrated eastwards by 10–50 km, also exposing the GRIP site to deformation close to the bed (Anandakrishnan *et al.* 1994). It appears likely that the GISP2 core is deformed from above the level at which Eemian ice first occurs, but Johnsen *et al.* (1995) argue that no large-scale overturn-folding appears to have disturbed the Eemian isotopic record in the GRIP core (because no two layers of exactly similar isotopic and chemical composition have been found). The GRIP core may therefore be a valid representation of climate change at least part-way into the Eemian above 2848 m (Johnsen *et al.* 1995). However, it is clear that additional work is required on the structural glaciology of the lowermost parts of both cores before the full climatic significance of the observed parameter variations can be assessed (Boulton 1993).

3. Abrupt environmental change and the ice core records

Evidence derived recently from both ice cores and marine sediments in Greenland and the North Atlantic region, respectively, suggests that marked climatic and other environmental changes have taken place over relatively short periods. The identification of layers of iceberg-rafted debris, 'Heinrich layers', in North Atlantic sediment cores at several times during the last 50 000 years is an example of the rapid collapse of the North American Laurentide Ice Sheet and the release of very large numbers of debris-laden icebergs into the North Atlantic (Bond *et al.* 1992). The duration of such events is on the order of 250–1000 years (MacAyeal 1993; Dowdeswell *et al.* 1995). However, it is the finer resolution provided by the ice-core record that allows the most detailed examination of the rapid nature of Late Quaternary climate change, demonstrating that a number of major climatic shifts in the North Atlantic region have occurred on timescales of clear relevance to human activities, sometimes as short as decades.

(a) The Younger Dryas

The time resolution of the GRIP and GISP2 ice cores has a relative accuracy of 1% or better over centuries at the depth of the Younger Dryas (about 1600 to 1700 m). The Younger Dryas, the last stadial before the Holocene, is bounded above and below by sharp changes in annual layer thickness, $\partial^{18}O$ values and dust content and is about 1300 years in duration (figure 3). The upper boundary is at 11 640 ± 250 or 11 550 ± 70 calendar years BP (i.e. before 1950), according the GISP2 and GRIP timescales, respectively (Alley *et al.* 1993; Johnsen *et al.* 1992). The Younger Dryas

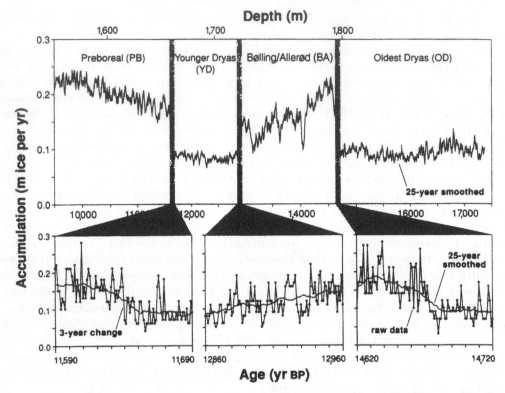

Figure 3. Accumulation rates in the GISP2 ice core between 10 000 and 17 000 years BP (from Alley *et al.* 1993). 25-year running means for accumulation are shown throughout the figure, with annual data superimposed in the lower three panels.

has more negative isotopic values, more dust and thinner ice layers than either the Preboreal above it or the underlying Bølling/Allerød interstadial (figure 3).

Alley *et al.* (1993) have emphasized the very abrupt termination of the Younger Dryas, pointing out that $\partial^{18}O$ values suggest termination over a period of about 50 years and that changing dust concentrations indicate a transition on the order of 20 years (Johnsen *et al.* 1992; Taylor *et al.* 1993). These rates confirm those seen in the Dye 3 core further south in Greenland by Dansgaard *et al.* (1989). However, detailed examination of changes in the thickness of annual ice layers, corrected for ice flow at depth, indicates a doubling of snow accumulation from the Younger Dryas to the Preboreal in as little as only three years (Alley *et al.* 1993). By a more conservative analysis, using 25 year running means for accumulation rates, accumulation doubled in 41 years. It is also pointed out that this very abrupt change is not a result of either enhanced shear-strain or thrust-faulting in this part of the ice core. The increase in accumulation rate between glacial conditions and the previous, Bølling/Allerød interstadial was also very abrupt, whereas the termination of that interstadial, marking the onset of the Younger Dryas, was less marked (figure 3).

Alley *et al.* (1993) suggest that the very rapidity of these shifts in environmental indicators places significant constraints on the nature of the climatic changes that can be responsible for them. Changes in accumulation may be linked to changing atmospheric dynamics, for example the duration or number of weather systems passing over Greenland, and to changes in the thermohaline circulation in the North Atlantic (Alley *et al.* 1993).

(b) The last glaciation

Isotopic values for the last glaciation switch between 24 intervals of relatively high and low $\partial^{18}O$ (figure 1), each lasting between several hundred to a few thousand years (Grootes *et al.* 1993). Variations are on the order of 4-6 per mille, which implies a temperature change of 7–8 °C. Full glacial conditions at the core sites were 10–13 °C colder than during the Holocene (Grootes *et al.* 1993). The intervals of less negative $\partial^{18}O$ values, representing warmer interstadials, lasted between about 500 and 2000 years (Johnsen *et al.* 1992). The isotopic records from the new central Greenland ice cores correlate well with similar evidence from other parts of the Greenland Ice Sheet, where shorter $\partial^{18}O$ time series for the last forty or so thousand years from Camp Century, Dye 3 and Renland also show an irregular series of stadials and interstadials (Johnsen *et al.* 1992).

An important element to this pattern of environmental change during glacial conditions back to about 100 000 years ago (figure 1) is that of the abruptness of the shift from cold to warmer temperatures at the onset of interstadials (Johnsen *et al.* 1992; Bond *et al.* 1993; Grootes *et al.* 1993). This shift may take place within as little as a few decades (Johnsen *et al.* 1992). By contrast, the initial part at least of the return to cooler stadial conditions is more gradual, sometimes followed by a more rapid fall to full stadial conditions (Grootes *et al.* 1993). The combination of repetitive interstadial and stadial patterns of isotopic variation has been termed a 'Dansgaard–Oeschger cycle', with each cycle having a characteristic saw-tooth form of very rapid warming to interstadials followed by slower cooling into the next stadial (Bond *et al.* 1993). These packages of warmer and colder oscillations within the last glacial, and in particular the sudden shift to warmer conditions, may be linked to the periodic rapid discharge of large numbers of icebergs into the North Atlantic (Bond *et al.* 1993), an issue that is discussed below.

(c) The Eemian Interglacial

If the $\partial^{18}O$ variations in the GRIP ice core during the last or Eemian interglacial represent a climatic proxy rather than the effects of ice deformation at depth, then several very rapid shifts in interglacial climate are implied within marine isotope Stage 5e (GRIP project members 1993). Substages 5e1, 5e3 and 5e5 are the warmest parts of the Eemian interglacial, and relatively unbroken intervals of warm climate within them lasted about 3000 years, less than 1000 years and 2000 years, respectively (figure 2). Each warm substage began and terminated with relative rapidity (on the order of decades), giving the Eemian record a variability quite unlike that observed in the very consistent $\partial^{18}O$ values from the Holocene (figure 1).

There are apparently several events of extreme rapidity within the interglacial record inferred from the GRIP core (figure 2). A catastrophic cooling lasting only about 70 years is located with Substage 5e1 (GRIP project members 1993). There are indications of an extremely rapid initiation of warming, followed by a less marked recovery, associated with this event. This is particularly apparent in the measurements of dust concentrations, which have been analysed with millimetre resolution and are affected less than other parameters by smoothing through diffusion at depth. There are also several major oscillations lasting up to about 750 years during the first 8000

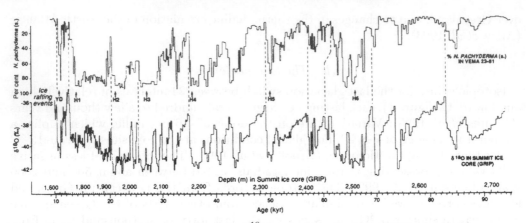

Figure 4. Comparison of the GRIP ice-core $\partial^{18}O$ record for the last 90 000 years with a proxy record for North Atlantic sea-surface temperatures and the occurrence of major ice-rafting events (adapted from Bond *et al.* 1993). Marine sediment core VEMA 23-81 is located in the North Atlantic west of Ireland and south of Iceland. Variations in the percentage of the foraminiferal species *N. pachyderma (s.)* correlate closely with sea-surface temperatures, and ice-rafting events are made up of Heinrich layers H-1 to H-6 and a further event during the Younger Dryas (YD).

years of the Eemian, where the cooling linked with each event of low $\partial^{18}O$ values is on the order of 10–14 °C (GRIP project members 1993).

Computer simulations using a global ocean model have mimicked these Eemian interglacial climatic states and the abrupt transitions between them through sudden shifts from one circulation pattern to another (Weaver & Hughes 1994). These sudden switches in the model ocean circulation were thought to be driven by a hydrological cycle which was assumed to be stronger during the Eemian interglacial than the Holocene, because the latter reached less high temperatures. However, two reservations are tied to this modelling (Broecker 1994); the first is that the Eemian environmental fluctuations in the GRIP core may simply be an artefact of the deformation of ice at depth and, secondly, the results produced by Weaver and Hughes' ocean model are likely to be heavily dependent on the surface boundary conditions that are used.

4. Ice cores and North Atlantic palaeoceanography

The causes of the observed variations in environmental parameters at the GRIP and GISP2 ice core sites in Greenland are often sought in the adjacent North Atlantic ocean. The good correlation between shifts in the thermohaline circulation of the North Atlantic during the last glacial period and the Greenland $\partial^{18}O$ record give evidence in support of this (Keigwin *et al.* 1994; McManus *et al.* 1994). The apparent link between pulses of rapid iceberg calving into the North Atlantic and the abrupt inception of warm interstadial conditions also suggests a strong ocean–atmosphere–cryosphere coupling (Bond *et al.* 1993).

(a) North Atlantic thermohaline circulation

The GRIP ice core record from Greenland has been compared with both the surface and deep-water circulation of the North Atlantic. Time series data on sea-surface temperatures and deep water circulation have been obtained through the analysis of planktonic and benthic foraminifera from marine sediment cores (Bond *et al.* 1993;

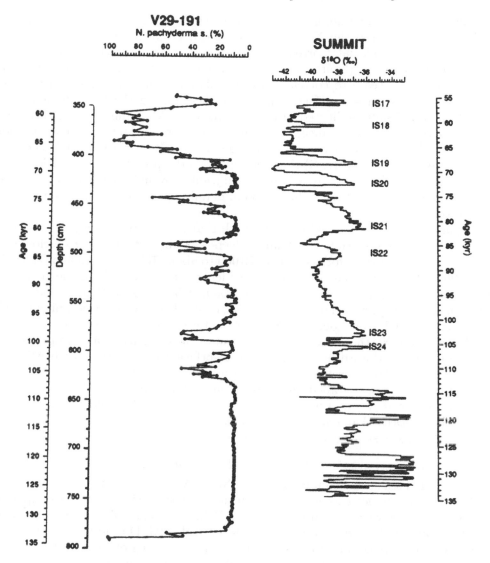

Figure 5. Comparison of the GRIP ice core record from Summit in Greenland with a proxy record for North Atlantic sea-surface temperatures for the period 60 000 to 135 000 years BP (adapted from McManus *et al.* 1994). Marine sediment core V23-191 is located in the North Atlantic west of Ireland and south of Iceland. Variations in the percentage of the foraminiferal species *N. pachyderma (s.)* correlate closely with sea-surface temperatures. Interstadials 17 to 24 are labelled. Note the lack of correlation for the last interglacial between the records, and the very low variability in *N. pachyderma (s.)* percentage.

Keigwin *et al.* 1994; McManus *et al.* 1994). Reconstructed sea-surface temperatures for the last 90 000 years show a series of warm-cold fluctuations on timescales of thousands of years, with particularly sharp terminations to the cold phases (figure 4). Bond *et al.* (1993) show that these fluctuations correlate closely with Dansgaard–Oeschger cycles in the ice core record, and suggest that the abrupt shift in ocean surface temperatures at the end of stadials is likely to represent a rise of several degrees over a few decades.

Using the same foraminiferal indicator of sea-surface temperatures, McManus *et al.* (1994) have analysed the North Atlantic record between 65–135 000 years ago, into the Eemian interglacial. Correlation with the GRIP ice core record is strong until about 105 000 years ago (figure 5), which is also about the time that the GRIP and GISP2 $\partial^{18}O$ records cease to correlate. Below this level, the marine sediment cores show a rather uniform sea-surface temperature field during the Eemian (figure 5), without the rapid shifts indicated in the GRIP ice core (McManus *et al.* 1994). This finding is of particular significance, since shifts in the North Atlantic ocean circulation pattern are often invoked as a cause of the observed $\partial^{18}O$ variations in Greenland. It implies one of the following: (i) that the GRIP record represents relatively localized climatic changes; (ii) that it is affected significantly by deformation, or (iii) that the ocean sediment records lack the temporal detail required to capture the Eemian climatic shifts.

Comparison of the GRIP $\partial^{18}O$ evidence with that from isotopic studies of benthic foraminifera from a western north Atlantic deep-sea sediment core shows a close correlation with ice core data for the period 70 000 to about 105 000 years ago (Keigwin *et al.* 1994). However, there is again no evidence for rapid shifts in deep marine isotopic parameters, which are a proxy for North Atlantic Deep Water production, during the Eemian interglacial (Keigwin *et al.* 1994). The benthic foraminiferal data reflect the shifting balance between North Atlantic Deep Water and Antarctic Bottom Water in the North Atlantic. This deep water flux is particularly important in maintaining the northward flux of heat in the surface waters of the North Atlantic. Keigwin *et al.* (1994) show that carbon isotope and carbonate concentration measurements reveal a more detailed signal than benthic foraminiferal $\partial^{18}O$ values. Using these parameters, they found no indication of variability in the deep ocean record of North Atlantic Deep Water in the Eemian, with the implication that variability in the GRIP ice core must have an alternative source.

(b) Heinrich events in the North Atlantic and rapid climate change

The record of sea-surface temperatures in North Atlantic cores not only correlates closely with the GRIP ice core time series during the last glaciation, but also appears to have clear links with ice rafted pulses or Heinrich events (figure 4). It appears that these rapid iceberg discharge events (MacAyeal 1993; Dowdeswell *et al.* 1995), and the lowered surface salinities associated with the melting of this ice (Bond *et al.* 1992), occurred during particularly cold stadial conditions (figure 4). The Heinrich events tend to occur, with the exception of H-1, close to the end of the cold period of Dansgaard–Oeschger cycles. The rapid warming, following the release of very large volumes of ice into the North Atlantic, may be associated with increasing surface salinities after the cessation of iceberg production. This increase, Bond *et al.* (1993) suggest, could be large enough to enhance the thermohaline circulation in the North Atlantic and the northward heat transfer that is associated with it.

5. Greenland ice cores and other Eemian climate records

The interglacial record of abrupt climate change derived from the GRIP core is not mirrored in the long Vostok ice core record from East Antarctica (Jouzel *et al.* 1993), which is the only existing ice core dataset of comparable length. However, this lack of correlation is not unexpected if it is assumed that interglacial fluctuations in climate at the summit of the Greenland Ice Sheet are linked to reorganizations of the oceanic and atmospheric circulation in the North Atlantic region, rather than to global factors. Indeed, the Antarctic ice core records respond only weakly to well-established climate changes in the North, such as the Dansgaard–Oeschger Events and the Younger Dryas cooling.

A problem in comparing the Greenland ice core records with marine and terrestrial evidence for climate change is that the latter are of lower time resolution and have a potentially long response time for changes in biotic systems (Larsen *et al.* 1995). Even so, there is little evidence in western European lowland pollen diagrams of the large-scale climatic fluctuations indicated in the last interglacial $\partial^{18}O$ record in the GRIP ice core. In the long Grande Pile peat bog record, for example, the Eemian appears as an interval of continuous warmth in Europe (Woillard 1978). By contrast, a number of marked fluctuations in pollen data are indicated for the last glaciation, although providing by no means as detailed a record as those from ice cores and North Atlantic sediments. This interpretation of a stable interglacial pollen record for the Eemian is supported by the continuous presence of species which are unable to withstand long periods of extreme winter cold, for example holly and ivy (Johnsen *et al.* 1995).

However, Johnsen *et al.* (1995) point out that palaeobotanical evidence from higher altitudes in Europe may provide a more sensitive record of interglacial environmental change, since climatic thresholds are more easily passed in such locations. Eemian records of magnetic susceptibility, pollen and organic carbon from lake deposits in the French Massif Central suggest two rapid cooling events during the interglacial, similar to those in the GRIP core (Thouveny *et al.* 1994).

It is concluded that the GRIP $\partial^{18}O$ record on the climate of the last interglacial has no immediate parallel in proxy environmental datasets from other areas (Johnsen *et al.* 1995). Three possible interpretations of this situation are given by Johnsen *et al.* First, the GRIP core may be deformed stratigraphically. Secondly, the evidence for environmental change at Summit may represent simply a climatic instability in a small portion of the north-western North Atlantic. Finally, interglacial climate changes may have taken place in a wider region, but are subdued and thus difficult to pinpoint using proxy indicators of limited temporal and environmental sensitivity.

6. Conclusions

The major changes which define climatic variability over the past million years are the glacial–interglacial cycles. Until recently, the existing paradigm was that such large changes, 5 °C or more shifts in average regional to global temperatures, occur over centuries to millennia, the timescale of orbital solar forcings. Recent evidence from ice cores has challenged this viewpoint. Glacial-interglacial scale changes in temperature, inferred from stable oxygen isotope ratios, have occurred in 50 years, or less than a human lifetime (figures 1 and 2). Changes in dust concentrations, which reflect dust source variability and wind strengths, can occur even faster, in 20 years. Shifts in deuterium excess, an isotopic parameter reflecting sea surface conditions, have also occurred in 20 years or less. Changes in the accumulation rate of snow, which reflect atmospheric temperature and storm frequencies, can occur faster still. A doubling of accumulation accompanying the end of the Younger Dryas occurred in only a few years (figure 3). These large, rapid changes are in contrast to the last 11 000 years which, in the Arctic, is a time of relatively stable climate (figure 1).

While the cause or causes of these changes is unknown, the speed implies a shift in atmospheric circulation, probably accompanied and amplified by changes in ocean circulation in the North Atlantic (figures 4 and 5). Suggested causes, such as periods of very rapid iceberg production, or Heinrich events, have usually relied on the existence of large amounts of ice on land, implying that rapid climate shifts are characteristic of glacial, and not of interglacial, periods. From an anthropogenic perspective, this is a fortuitous situation. For this reason, the results from the GRIP ice core, with its highly variable record of Eemian interglacial climate (figure 2), are particularly important. While this result has not been duplicated in another ice core, particularly not in the GISP2 core which is only 28 km away from the GRIP site, neither has there been conclusive evidence that the GRIP record is fatally flawed by folding of the ice. Other evidence from ocean sediment records (figure 5) and land-based pollen records during the Eemian have yielded mixed results, with the current weight of evidence perhaps falling on the side of a stable Eemian. It would be unwise to draw conclusions about the variability of interglacial climate, however. The temporal detail available in ice cores is difficult to obtain in other sedimentary records, and another ice core from Greenland may be required to resolve the issue. Until then, as human activity continues to modify the atmospheric system, possible future scenarios for climate change should include one which incorporates major shifts on timescales of years to decades.

The identification of abrupt climate change in the polar regions is dependent on a large number of data sources and measurements, many of them undertaken in difficult field conditions. We thank all of the investigators who have dedicated their time to collecting, analysing and interpreting the thousands of observations reviewed here.

References

Alley, R. B., Meese, D. A., Shuman, C. A., Gow, A. J., Taylor, K. C., Grootes, P. M., White, J. W. C., Ram, M., Waddington, E. D., Mayewski, P. A. & Zielinski, G. A. 1993 Abrupt increase in Greenland snow accumulation at the end of the Younger Dryas event. *Nature, Lond.* **362**, 527–529.

Anandakrishnan, S., Alley, R. B. & Waddington, E. D. 1994 Sensitivity of the ice-divide position in Greenland to climate change. *Geophys. Res. Lett.* **21**, 441–444.

Bond, G. *et al.* 1992 Evidence for massive discharges of icebergs into the North Atlantic ocean during the last glacial period. *Nature, Lond.* **360**, 245–249.

Bond, G., Broecker, W., Johnsen, S., McManus, J., Labeyrie, L., Jouzel, J. & Bonani, G. 1993 Correlations between climate records from North Atlantic sediments and Greenland ice. *Nature, Lond.* **365**, 143–147.

Boulton, G. S. 1993 Two cores are better than one. *Nature, Lond.* **366**, 507–508.

Broecker, W. S. 1994 An unstable superconveyor. *Nature, Lond.* **367**, 414–415.

Dansgaard, W. *et al.* 1993 Evidence for general instability of past climate from a 250-kyr ice-core record. *Nature, Lond.* **364**, 218–220.

Dansgaard, W., White, J. W. C. & Johnsen, S. J. 1989 The abrupt termination of the Younger Dryas. *Nature, Lond.* **339**, 532–534.

Dowdeswell, J. A., Maslin, M. A., Andrews, J. T. & McCave, I. N. 1995 Iceberg production, debris rafting, and the extent and thickness of Heinrich layers (H-1, H-2) in North Atlantic sediments. *Geology* **23**, 301–304.

Greenland ice-core project (GRIP) members 1993 Climate instability during the last interglacial period recorded in the GRIP ice core. *Nature, Lond.* **364**, 203–207.

Grootes, P. M., Stuiver, M., White, J. W. C., Johnsen, S. J. & Jouzel, J. 1993 Comparison of oxygen isotope records from the GISP2 and GRIP Greenland ice cores. *Nature, Lond.* **366**, 552–554.

Johnsen, S. J., Dansgaard, W. & White, J. W. C. 1989 The origin of Arctic precipitation under present and glacial conditions. *Tellus* **B41**, 452–468.

Johnsen, S. J., Clausen, H. B., Dansgaard, W., Fuher, K., Gundestrup, N., Hammer, C. U., Iversen, P., Jouzel, J., Stauffer, B. & Steffensen, J. P. 1992 Irregular glacial interstadials recorded in a new Greenland ice core. *Nature, Lond.* **359**, 311–313.

Johnsen, S. J., Clausen, H. B., Dansgaard, W., Gundestrup, N. S., Hammer, C. U. & Tauber, H. 1995 The Eem stable isotope record along the GRIP ice core and its interpretation. *Quatern. Res.* **43**, 117–124.

Jouzel, J. *et al.* 1993 Extending the Vostok ice-core record of palaeoclimate to the penultimate glacial period. *Nature, Lond.* **364**, 407–412.

Keigwin, L. D., Curry, W. B., Lehman, S. J. & Johnsen, S. 1994 The role of the deep ocean in North Atlantic climate change between 70 and 130 kyr ago. *Nature, Lond.* **371**, 323–326.

Larsen, E., Sejrup, H. P., Johnsen, S. J. & Knudsen, K. L. 1995 Do Greenland ice cores reflect NW European interglacial climate variations? *Quatern. Res.* **43**, 125–132.

MacAyeal, D. R. 1993 Binge/purge oscillations of the Laurentide Ice Sheet as a cause of the North Atlantic's Heinrich Events. *Paleoceanography* **8**, 775–784.

McManus, J. F., Bond, G. C., Broecker, W. S., Johnsen, S., Labeyrie, L. & Higgins, S. 1994 High-resolution climate records from the North Atlantic during the last interglacial. *Nature, Lond.* **371**, 326–329.

Taylor, K. C., Hammer, C. U., Alley, R. B., Clausen, H. B., Dahl-Jensen, D., Gow, A. J., Gundestrup, N. S., Kipfstuhl, J., Moore, J. C. & Waddington, E. D. 1993 Electrical conductivity measurements from the GISP2 and GRIP Greenland ice cores. *Nature, Lond.* **366**, 549–552.

Thouveny, N. *et al.* 1994 Climate variations in Europe over the past 140 kyr deduced from rock magnetism. *Nature, Lond.* **371**, 503–506.

Weaver, A. J. & Hughes, T. M. C. 1994 Rapid interglacial climate fluctuations driven by North Atlantic ocean circulation. *Nature, Lond.* **367**, 447–450.

Woillard, G. M. 1978 Grande Pile peat bog: a continuous pollen record for the last 140 000 years. *Quatern. Res.* **9**, 1–21.

Non-steady behaviour in the Cenozoic polar North Atlantic system: the onset and variability of Northern Hemisphere glaciations

By J. Thiede[1] and A. M. Myhre[2]

[1] GEOMAR Research Center for Marine Geosciences,
Kiel University, 24148 Kiel, Germany
[2] Department of Geology, University of Oslo, 0316 Oslo, Norway

Changes of the extent of the Arctic Ocean sea-ice cover over the past century, the geological record of the Arctic Ocean seafloor of the youngest geological past, as well as the evidence of a pre-Glacial temperate to warm Arctic Ocean during Mesozoic and Palaeogene time are witnesses of dramatic revolutions of the Arctic oceanography. The climate over northwestern Europe on a regional scale as well as the global environment have responded to these revolutions instantly over geological time scales. Results of ocean drilling in the deep northern North Atlantic document an onset of Northern Hemisphere glaciation towards the end of the middle Miocene (10–14 Ma). While the available evidence points to early glaciations of modest extent and intensity centred around southern Greenland, the early to mid-Pliocene intervals record a sudden intensification of ice-rafting in the Labrador and Norwegian Greenland seas as well as in the Arctic Ocean proper. The Greenland ice cap seems to have remained rather stable whereas the northwest European ice shields have experienced rapid and dramatic changes leading to their frequent complete destruction. Many sediment properties seem to suggest that orbital parameters (Milankovitch-frequencies) and their temporal variability control important properties of the deep sea floor depositional environment. Obliquity (with approximately 40 ka) seems to be dominant in pre-Glacial (middle Miocene) as well as Glacial (post late Miocene) scenarios whereas eccentricity (with approximately 100 ka) only dominated the past 600–800 ka. Plio-Pleistocene deposits of the Arctic Ocean proper, of the entire Norwegian Greenland and of the Labrador seas have recorded the almost continuous presence of sea-ice cover with only short 'interglacial' intervals when the eastern Norwegian Sea was ice-free. The documentation of long-term changes of the oceanographic and climatic properties of the Arctic environments recorded in the sediment cover of the deep-sea floors might also serve to explain scenarios which have no modern analog, but which might well develop in the future under the influence of the anthropogenic drift towards warmer global climates.

1. Introduction

The Arctic and sub-Arctic seas exert major influences on global climate and ocean systems. High northern latitude oceans directly influence the global environment through the formation of permanent and seasonal ice covers, transfer of sensible

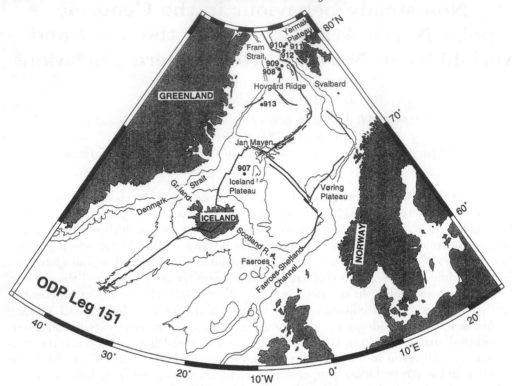

Figure 1. Bathymetric map of the Norwegian Greenland Sea, showing drill site locations for
Leg 151. Contour interval is 2000 m.

and latent heat to the atmosphere, deep-water renewal and deep-ocean ventilation.
Thus, any serious attempt to model and understand the Cenozoic variability of global
climate must take into account the palaeoenvironmental changes in the Arctic and
sub-Arctic deep-sea basins.

Due to the permanent sea-ice cover, the Arctic Ocean is the only major ocean basin
that has not been drilled by DSDP or ODP. Of all the Arctic and sub-Arctic deep-sea
basins the Norwegian Greenland Sea (figure 1) without doubt has the most important
influence on evolution and change of the Northern Hemisphere palaeoenvironments;
therefore the arguments presented in this paper will mainly be drawn from the ODP
and DSDP legs devoted to this area.

2. Cenozoic climate evolution of high northern latitudes

It is still not known how and when climatic, tectonic, and oceanographic changes
in the Arctic (Thiede *et al.* 1992) contributed to Cenozoic global ocean cooling and
to increased thermal gradients. In order to understand the Cenozoic evolution of
the global climate system, it is necessary to clarify when the Arctic Ocean became
ice-covered and to document the variability of ice covers in the Arctic. It has been
proposed that the Arctic Ocean has been permanently ice-covered since the beginning
of the late Miocene (Clark 1982) or even earlier (Wolf & Thiede 1991). Other studies
concluded that this event happened in the Matuyama or at the Brunhes/Matuyama
boundary (Herman & Hopkins 1980; Carter *et al.* 1986; Repenning *et al.* 1987).

A major threshold of the climate system was passed with the inception of glaciers and ice sheets in the Northern Hemisphere. Data from ODP Leg 104 document minor input of ice-rafted debris (IRD) into the Norwegian Sea in the late Miocene and through the Pliocene, pointing to the existence of periods when large glaciers were able to form and reach coastal areas in some of the regions surrounding the northernmost North Atlantic (Jansen & Sjøholm 1991; Wolf & Thiede 1991). IRD data from ODP Leg 105, Site 646, suggest the onset and discontinuous early existence of ice in the Labrador Sea to the south of Greenland since middle/late Miocene times (Wolf & Thiede 1991). The major shift to a mode of variation characterized by repeated large glacials in Scandinavia probably occurred at about 2.5 Ma† and was further amplified at about 1 Ma (Jansen *et al.* 1988; Jansen & Sjøholm 1991). Terrestrial data indicate significant cooling on Iceland at about 10 Ma (Mudie & Helgason 1983) and glaciation in elevated areas of Iceland in the latest Miocene and the Pliocene (Einarsson & Albertsson 1988). Terrestrial evidence also suggests forested areas in the Arctic fringes, which are far north of the present forest/tundra boundary, until about 2 Ma (Carter *et al.* 1986; Nelson & Carter 1985; Funder *et al.* 1985; Repenning *et al.* 1987). The chronology from these land sites is, however, poorly constrained, and there are no continuous records from land sites that document the climatic transition into a cold Arctic climate.

In addition to the above questions that address the magnitude of glaciations and the passing of certain climatic thresholds in the Earth's history, the frequency components of the climatic, oceanographic, and glacial evolution of the Arctic and sub-Arctic are of importance for assessing the climate system's response to external forcing. Results from DSDP Leg 94 sites in the North Atlantic have shown that sea-surface temperatures and ice volumes have a strong response to orbital forcing over the last 3 Ma. However, the amplitudes of climatic changes and the dominant frequencies have varied strongly, indicating variations in the way the climate system responds to external forcing (Ruddiman *et al.* 1986; Ruddiman & Raymo 1988; Raymo *et al.* 1990). Work is under way, based on Leg 104 material (Jansen & Sjøholm 1991; Henrich 1992) to study the cyclicity of IRD input into the sub-Arctic Norwegian Sea. This can aid in understanding the controlling factors for subpolar ice-sheet variations. However, available data do not permit extending this type of high-resolution study on orbital time scales to other parts of the Arctic Ocean and the Norwegian Greenland Sea.

3. Environmental frame of the Norwegian Greenland Sea

The Cenozoic history of the boundary between the North American and Eurasian plates (figure 1) has generated a series of interconnected deep-sea basins which exchange their water masses through the North Atlantic-Arctic gateways. The basins consist of the main North Atlantic in the south, and the Norwegian Greenland Sea and the eastern Arctic basin in the north. Water exchange between them is constricted by two gateways, namely in the south by the Greenland–Scotland Ridge with Denmark Strait in the west and the Faroe–Shetland Channel in the east, and in the North by Fram Strait between the continental margins of northeastern Greenland and Svalbard. The geologically young connection between the Atlantic and Arctic oceans has proven of great importance because they allowed climatically highly sensitive surface and bottom currents to develop. To study how the depositional environment in and around the two North Atlantic Arctic gateways responded to the

† 1 Ma = 1 million years, 1 ka = 1000 years

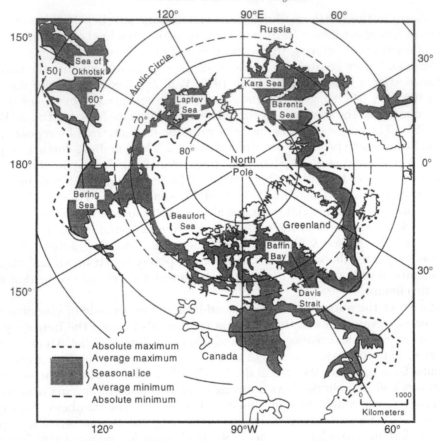

Figure 2. Average and extreme sea ice limits (greater than $\frac{1}{8}$), after CIA (1978) and Barry (1989).

Cenozoic tectonic and climatic changes in the area, was the main scientific objective of ODP Leg 151, the first of two drilling legs devoted to the North Atlantic-Arctic gateway problem.

The eastern segment of the Greenland–Scotland Ridge is crossed by the North Atlantic Drift (figure 2) which imports temperate surface water masses into the Norwegian Sea. After branching into the North Sea these waters, now called the Norwegian Current, follow the Norwegian continental margin bounded to the east by the Norwegian Coastal Current, to the west by the Arctic polar water masses of the central Norwegian Greenland Sea. Off northern Norway the Norwegian Current branches into the North Cape Current, turning to the east and the West Spitsbergen Current, trailing the Barents Sea continental margin to the north until it dips below the Arctic sea-ice cover in the northern part of Fram Strait. In the west these inflowing Atlantic waters are partly balanced by the East Greenland Current which carries cold, brackish, partly and seasonally highly variable ice-covered Arctic surface waters from the Arctic Ocean along the East Greenland continental margin into the northwestern Atlantic Ocean.

The deep waters of the Norwegian Greenland Sea are renewed by two principally different processes which both produce very cold, saline, dense and oxygenated bottom waters, namely the down-welling of surface waters between the Polar and

ODP LEG 151 Sites

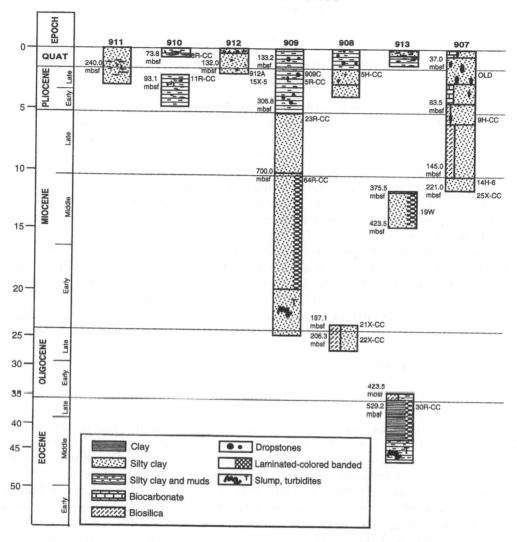

Figure 3. Lithologic summaries of Leg 151 drill sites, showing lithologic units and ages. (From Myhre *et al.* 1995.)

Arctic fronts in the open Greenland Sea (Koltermann 1987) and as a result of seasonally highly variable sea-ice formation by the production of dense brines on the shelves which then flow across the continental margins into the adjacent deep-sea basins (Quadfasel *et al.* 1987). The Norwegian Greenland Sea is filled beyond the sill depth of the southern gateway by these cold deep waters which spill through Denmark Strait, the Iceland Faroe Ridge and the Faroe–Shetland Channel into the deep North Atlantic. The North Atlantic deep water (NADW) is fed by these water masses as they flow across various segments of the Greenland–Scotland Ridge. The overflow of cold, saline and oxygenated deep waters from the Norwegian Greenland Sea into the North Atlantic Ocean (Meincke 1983) and its historic variability as documented in the ocean sediments is one of the most dynamic processes of the global environment (Broecker 1991) and is expected to respond sensitively to any climatic change of the high northern latitudes.

4. Previous ocean drilling in Northern Hemisphere polar and subpolar deep-sea basins

A history of scientific drilling in the Northern Hemisphere polar and subpolar deep-sea basins for studying the palaeoenvironment and Cenozoic palaeoclimate would not be complete without a brief excursion to the North Pacific Ocean DSDP Leg 18 and 19 results (Kulm *et al.* 1973; Creager *et al.* 1973). Both legs are part of the very early DSDP activities, but they visited the northern rim of the Pacific and the Bering Sea, providing some important data on the imprint of the onset of Northern Hemisphere Cenozoic glaciations in an area geographically opposite to the Norwegian Greenland Sea drill sites. DSDP Leg 18 visited the northern Pacific and found glacio-marine deposits in Sites 178–182 (Alaska Abyssal Plain, Aleutian Trench and continental margin off southwest Alaska). The oldest record of ice-rafted erratics has been observed in the glacio-marine deposits of the Alaska Abyssal Plain, where upper Pliocene and Pleistocene deposits down to 258 mbsf (metres below sea-floor) with erratics can be subdivided into 3 lithologic units, each with variable amounts of erratics and henceforth documenting a certain pattern of temporal variability of ice-rafting. No attempt has been made to identify a potential North American source region.

DSDP Leg 19 continued the programme of Leg 18 towards the West, but it also crossed over the Aleutian island chain into the deep Bering Sea. Like Leg 18, it stayed in relatively low latitudes (south of 57° N) and therefore the drill sites are not well placed to address onset and evolution of Cenozoic Northern Hemisphere glaciations. The possibility of making pertinent observations was further reduced by the fact that these early legs only performed spot rotary-coring, obtaining an incomplete and intensely disturbed record of the youngest parts of the sedimentary sequences. However, evidence for ice-rafting with variable intensity has been found in sediments possibly as old as early Pliocene (Site 187), almost always in the Upper Pliocene to Quaternary deposits at most of the drill sites. No attempts have been made to quantify ice-rafting or to determine potential provenances of the erratics. Later, ODP Leg 145 also visited the North Pacific Ocean and continuously cored several deep sites which showed that the onset of the Northern Hemisphere glaciation occurred at 2.6 Ma, marked by abundant dropstones coming both from Siberian and Alaskan sources (Rea *et al.* 1993*b*).

DSDP visited the North Atlantic several times, but only once the Norwegian Greenland Sea (Talwani *et al.* 1976) during Leg 38. Spot- and rotary-coring and mostly geophysical and tectonic objectives did not allow the deciphering of much of the detail of the Late Cenozoic palaeoenvironmental history. Many of the Norwegian Greenland Sea drill sites contained ice-rafted detritus (both erratics and Cretaceous *Inoceramus* prisms are mentioned); their stratigraphic distribution confirmed that sea-ice covers and ice-rafting were not restricted to the Quaternary. Ice-rafting exhibiting considerable temporal variation extended clearly into the Tertiary, at that time Pliocene. Warnke & Hansen (1977) based on the Leg 38 material, later confirmed this and established a regional distribution of ice-rafting with maxima along the Greenland and Norwegian continental margin.

ODP Leg 104 was the next major contribution towards deciphering the history of Northern Hemisphere Cenozoic cooling and glaciation. The major scientific objective of the central drilling location on the Vøring Plateau off mid-Norway was oriented towards sampling a thick, dipping reflector sequence of volcanic origin (Eldholm *et al.* 1987) related to the initial opening of the Norwegian Greenland Sea (Hinz *et al.* 1993; Eldholm & Thomas 1993). However, together with additional drill sites on either side along a transect across the Vøring continental margin, this site also revealed important data about the history of the Norwegian Current and the onset of Northern Hemisphere Cenozoic glaciations.

5. New deep-ocean drilling in high northern latitudes: Leg 151

Following recommendations defined after Leg 104 (Thiede *et al.* 1989) and further specified by the NAAG DPG (Ruddiman *et al.* 1991), Leg 151 was scheduled to drill a series of sites (figure 1) in several remote geographic, partly ice-covered locations (the northern gateway region, i.e. Yermak Plateau and Fram Strait, the East Greenland Margin, and the Greenland-Norway Transect, the Iceland Plateau, the Greenland–Scotland Ridge) with the aim of reconstructing the temporal and spatial variability of the oceanic heat budget and the record of variability in the chemical composition of the ocean. Leg 151 was also to undertake a study of circulation patterns in a pre-glacial, relatively warm polar and subpolar ocean, and the mechanisms of climatic change in a predominantly ice-free climatic system. In addition, the proposed drilling included a collection of sequences containing records of biogenic fluxes ($CaCO_3$, opal and organic carbon) and stable-isotopic carbon and oxygen records which addressed aspects of facies evolution and depositional environments as well as the carbon cycle and productivity. The drilling approach focused on rapidly deposited sediment sequences to be used for high-resolution, Milankovitch scale palaeoclimatic analysis and rapid sub-Milankovitch-scale climate changes.

The voyage of the Joides Resolution during Leg 151 to the Norwegian Greenland–Iceland Sea, Fram Strait and the marginal Arctic Ocean, brought a scientific drilling vessel into higher northern latitudes with waters more ice-infested than ever before. The drill ship was escorted by the Finnish icebreaker Fennica. ODP Leg 151 opened completely new scientific perspectives to Arctic geoscientific research, thus representing a historic step in the scientific exploration of the Arctic. The sedimentary sections penetrated at these sites were investigated to unravel the history of surface and bottom waters in the Norwegian Greenland Sea and in the Arctic Ocean. These are connected through the narrow northernmost gateway of the Atlantic Ocean, the Fram Strait, with a complicated history of water exchange between these two polar to subpolar Northern Hemisphere deep-sea basins. Yermak Plateau and Fram Strait are also relatively young geological features whose origin is not known in enough detail to resolve their impact on the changes in current patterns and whose basement age and nature as well as subsidence have to be established by means of deep-ocean drilling. Four areas of the North Atlantic–Arctic Gateway province have been drilled during ODP Leg 151 during July to September 1993 (figure 1):

(i) Site 907, located on the eastern-central Iceland Plateau, was drilled to obtain Quaternary and Neogene biogenic and terrigenous sediment sequences with a detailed palaeoenvironmental record;

(ii) Sites 908 and 909, in the southern Fram Strait, represent a depth transect with shallow (on Hovgård Ridge) and deep (on the Greenland Sea/Arctic Ocean sill depth) locations to study the history of the water exchange between the two adjacent polar deep-sea basins as well as the opening and tectonic history of Fram Strait;

(iii) a suite of sites (Sites 910, 911 and 912) was planned for the Yermak Plateau to study the origin of its basement as well as its history of subsidence, and to establish as far back as possible the history of the truly Arctic marine palaeoenvironment along a depth transect as well as its interaction with the Upper Cenozoic Arctic ice sheets; and

(iv) sites at the East Greenland Margin (Site 913) were aimed at studying the history of the East Greenland Current and of the Greenland Sea.

The main scientific objectives of ODP Leg 151 comprised the investigation of the short and long-term variability of the Late Cenozoic palaeoceanography of the Norwegian Greenland Sea and of the Arctic Ocean proper, which is controlled by or responds to tectonic evolution and Northern Hemisphere palaeoclimatic change. It was important to investigate properties and changes of the surface and bottom water masses, as well as the history of their main current systems. Special attention was paid to the history of the sea-ice cover as well as to glaciations and deglaciations on the adjacent continents. Following ODP Leg 104, which addressed the processes controlling the origin of the dipping reflector sequence on the Vøring Plateau and the history of the temperate Norwegian Current in the eastern Norwegian Greenland Sea, the cluster of drill sites at ODP Leg 151 combines shallow and deep locations in the western and northern Norwegian Greenland and in the Arctic Ocean proper (figure 1). Detailed descriptions of the stratigraphic sections penetrated at each of the drill sites have recently been published in the Initial Reports of ODP Leg 151 (Myhre *et al.* 1995). We therefore present here only a brief interpretation of the palaeoenvironmental evolution of the northern part of the North Atlantic Arctic gateway as it can be deduced from said drilling results.

6. Narrative of Cenozoic palaeoenvironmental evolution of high Northern Hemisphere deep-sea basins

(a) Middle Eocene (Site 913)

The lowermost sediments at Site 913, the oldest material recovered during Leg 151, contain multiple-fining upward-sediment gravity flows with coal and mud clasts, and laminations, with highest abundances of terrigenous organic matter recovered during Leg 151. This suggests a site close to a continental source, possibly in an active tectonic or high-sedimentation-rate area, such as an early post-rift basin.

Above the mass-wasted sediment are finer-grained, interbedded, laminated and massive deposits, with moderate bioturbation. These sediments show a general fining upward trend in the middle Eocene, suggesting a change in sediment source, depocentre, or possibly a tectonic barrier, preventing coarse sediment influx. The chemistry of the sediment shows little change, consisting of high silica continental sediments, derived from a granitic source. High terrigenous organic carbon values continue, although they also decrease through the middle Eocene. The abundance of fish teeth found in some samples from this interval suggests a slow sedimentation rate.

Palaeontological evidence from Site 913 suggests increasing productivity throughout the middle Eocene. Biogenic silica is preserved only in the upper middle Eocene. Decalcified *Bolboforma* and the absence of calcareous planktonic and benthic foraminifers and nannofossils place this site below the calcite compensation depth (CCD).

Similarities to the agglutinated benthic foraminifers in the Labrador Sea suggest that the bottom waters were in connection with the North Atlantic.

(b) Late Eocene to early early Oligocene (Site 913)

At Site 913 there is a renewed influx of terrigenous organic carbon in the late Eocene, at approximately the same level as the first appearance of biogenic silica. The sediments themselves show little change, being interbedded, massive and laminated silty clay and clay. Up-section, however, they become very colourful, with exciting shades of blue, purple and green. This coincides with an increase in the preservation and abundance of siliceous microfossils. Discrete siliceous ooze intervals were recovered and SiO_2 is also high in the pore waters. The siliceous intervals are formed during times of high productivity, resulting in high pelagic sedimentation rates and high values of organic carbon.

The high productivity is believed to have been caused by upwelling. Decay of organic matter would result in low oxygen conditions, decreased bottom-water pH and the dissolution of carbonate. As a result calcareous microfossils are absent during this interval. Similarities to the agglutinated benthic foraminifers in the Labrador Sea suggest that the bottom waters were exchanged with the North Atlantic.

(c) Late Oligocene to early early Miocene (Sites 908, 909?)

Evidence for this interval from Site 908 suggests moderately well-mixed oceanic conditions in the Norwegian Greenland Sea. The predominantly fine-grained and hemipelagic sediments record relatively high, but fluctuating, surface-water productivity. This is particularly well demonstrated by organic carbon concentrations showing the highest variability of any Leg 151 site, generally between 0.75% and 1.5%, although some layers were over 2%. The average values are also higher than at most other Leg 151 sites. High productivity is further supported by the abundance of siliceous microfossils, with a diverse assemblage of diatoms and a low diversity assemblage of radiolarians. Absence of planktonic foraminifers, rare nannofossils, and benthic foraminifers place this site below the calcite lysocline during this time. Intermediate bottom-water oxygen content is suggested by the benthic foraminiferal morphologies and the low diversity of the assemblage. Extensive bioturbation suggests at least intermediate oxygenated bottom waters, although thin, poorly bioturbated, laminated intervals suggest fluctuations to lower oxygen levels.

(d) Middle Miocene

A glimpse of the middle Miocene is observed in the laminated clay and sandy muds of Core 913-19W which contains a moderate abundance of siliceous microfossils with low-diversity radiolarians and moderate diversity diatoms. Ebridians are also common. This suggests moderately high pelagic productivity, although the organic carbon values are very low, less than 0.5%. Dissolution of all calcareous biota and the occurrence of a low diversity assemblage of siliceous agglutinated benthic foraminifers suggest that this site was below the CCD during this time.

(e) Early to late Miocene and early Pliocene

This interval is recovered in two dramatically different sections in Site 909 (Fram Strait) and 907 (Iceland Plateau). In the southern Norwegian Greenland Sea, Site 907 suggests high productivity of siliceous microfossils, although preserved organic carbon is low, less than 0.5%. Terrigenous input is also low, and volcanic glass forms about 10% of the sediment. Diatoms suggest upwelling conditions and an Atlantic source of surface water. The resulting lowered pH causes the dissolution of all carbonate. These oceanographic conditions are interrupted during two brief intervals with enhanced preservation of carbonate, and deposition of the only nannofossil ooze recovered during Leg 151. Timing of these events is constrained by the calcareous microfossils and palaeomagnetics.

At Site 909, in the Fram Strait, the lower Miocene to Pliocene interval suggests restricted oceanic circulation. On average, total organic carbon is slightly higher than 1%. The sediments show a general fining upward trend. They alternate on a decimetre scale between massive, moderately to extensively bioturbated, and laminated, weakly to unbioturbated sediments, probably reflecting changing bottom water conditions. No biogenic silica or carbonate is preserved. Agglutinated benthic foraminifers record very low oxygen bottom water in the late Miocene to early Pliocene, supporting the interpretation of corrosive bottom water. The lowermost sediments at Site 909 contain a few Oligocene nannofossils.

The upper Oligocene to lower Miocene section at Site 909 differs considerably from Site 908, which presents a problem, since they are adjacent though at different depths. Possibly, at Site 908 there are undetected nonconformities which can be invoked to explain these differences. Likewise, the Miocene section at Site 913 is very different, but geographically further away. The laminations here, as with the laminated middle Eocene section at Site 913, may have been produced when the basin circulation was restricted.

The oldest dropstone recovered on Leg 151 is from the uppermost Miocene at Site 907. This is somewhat surprising because this site also has the lowest abundance of dropstones. The next oldest dropstone is also an isolated occurrence, this time in Site 909, however, its location at the edge of a drilling biscuit makes it suspect.

(f) Pliocene to Quaternary

Pliocene and Quaternary sediments were recovered from all sites, (Sites 910, 911, 912: Yermak Plateau, Sites 909 and 908: Fram Strait, Site 913: East Greenland Margin, and Site 907: Iceland Plateau), and are dominated by silty clay and muds. With the exception of Site 907, the first occurrence of dropstones at all sites, based on palaeomagnetic ages, is in the lower Pliocene. At several sites, e.g. Site 908, 909, 911, the onset of dropstones is preceded by an increase in potash (K_2O) concentration in the sediment, suggesting a change in the sediment source. The introduction of fresher, less weathered sediment appears to proceed the onset of glacial dropstones. The increasing potassium trend continues as the abundance of dropstones increases, and then levels off at the higher value.

At about 2.5 Ma most sites show a substantial increase in dropstone abundance. Above this the number of dropstones at each site (normalized for recovery) shows different absolute abundances and types of fluctuations. In most sites, there is a substantial decrease in dropstone abundance in the upper Quaternary. Unfortunately the age resolution, and the location of nonconformities are not well constrained at this time, and hence it is not possible to see if these changes reflect glacial/interglacial changes.

Microfossils are generally more common in this interval, more so in the youngest part than in the older sections. Pre-glacial fossiliferous Pliocene sediments were recovered only at Site 910. The well preserved carbonate microfossil groups record good circulation and carbonate preservation at this shallow site. Organic carbon values average about one percent. High glacial, especially Quaternary, sedimentation rates (greater than $10 \, cm \, ka^{-1}$) are recorded at the Yermak Plateau Sites 911 and 912, whereas Site 910, also on the Yermak Plateau, has a very high Pliocene sedimentation rate.

At least four sites (907, 909, 911 and 912) record a lower Quaternary to upper Pliocene calcium carbonate dissolution event. This event was also observed on the Vøring Plateau. Dissolution of carbonate is caused by increased CO_2 in the bottom water. This occurs during ageing, for instance when bottom-water formation decreases.

The presence of biogenic sediments, particularly in the upper section (less than 50 mbsf) at most sites, suggests more mild climatic conditions. These sediments are more colourful, with olive gray and very dark gray layers.

At Sites 908, 910 and 911, there is a steep increase in the bulk density and strength over the top 20 to 30 mbsf (upper Quaternary), followed by a gradual decrease with depth. These are three of the four shallowest sites (1273, 556, and 901 mbsf respectively). Data at the other shallow site, Site 912 (1037 mbsf), is not good enough to see any trend. Although no obvious compositional changes were observed, the anomalously high values may be related to overcompaction due to ice, or may possibly be a permafrost feature due to freezing. This problem will be investigated in considerably more detail when cores from the geotechnical hole, Hole 910D are studied. Physical properties such as bulk density of sections older than 600–700 ka show remarkably regular fluctuation with time, imaging the 40 ka-obliquity Milankovitch frequency.

7. Future Northern Hemisphere high latitude deep ocean drilling

After the successful North Atlantic legs of the Ocean Drilling Program (Legs 104, 105, 151, 152, and the upcoming second NAAG-leg (ODP Leg 163 in 1995) and the North Pacific Transect, which revisited (Rea *et al.* 1993a) areas that were drilled during DSDP Legs 18 and 19, the question remains if and how deep-sea drilling can contribute in the future to solving the mysteries of the geological, environmental and biological history of the Northern Hemisphere polar and subpolar deep-sea basins. Once the second NAAG-Leg with its important scientific targets in the southern Norwegian Greenland Sea and on both sides of the Greenland–Scotland Ridge is completed, many of the presently defined high priority drilling targets will have been exhausted.

Considering only the available and established drilling techniques the following future high priority areas can be suggested to further study the Northern Hemisphere palaeoenvironment.

(i) *Bering Sea.* Transition from a pre-glacial to a glacial palaeoenvironment, variability of the glacial palaeoceanography, back-arc-spreading phenomena.

(ii) *Sea of Okhotsk.* Late Cenozoic palaeoenvironments, areas of high fluid and gas exchange from the sea floor into the overlying water column (areas of highest methane emission, seeps and their geological setting and history), back-arc spreading phenomena, Kamtchatka volcanic and tectonic history.

(iii) *North Atlantic–Arctic.* Deep penetration on Yermak-Plateau to establish nature and age of basement as well as of the overlying sediment column. Transect across one of the trough-mouth fans with the aim of correlation to the glacial history of the continental hinterland.

Considering alternative drilling platforms the plans of the Nansen Arctic Drilling

(NAD) programme (Thiede & NAD Scientific Committee 1992) receive the highest priority. It is the aim of this programme to bring deep-sea drilling into the permanently ice-covered Arctic Ocean to resolve its tectonic and palaeoenvironmental history. Areas of high priority have been established as follows.

(i) *Alpha–Mendeleev Ridge*. Sampling of the Mesozoic and lower Cenozoic preglacial pelagic sedimentary sequence of the Arctic Ocean, establishment of the age and nature of the volcanic basement.

(ii) *Lomonosov Ridge*. Sampling of extended post-rift sedimentary sequence on top of the Lomonosov Ridge to establish timing of onset of the Arctic Ocean glaciation and of the variability of the depositional environments.

(iii) *Laptev Sea*. Intersection of the active mid-ocean Gakkel Ridge with the Laptev Sea continental margins, tectonic history of the area, nature of rifting, palaeoenvironment of high resolution terrigenous sections in front of a large Arctic delta.

In the preparation of this paper we have been able to draw on the latest version of the Leg 151 Initial Results (Myhre *et al.* 1995), the site proposals submitted by Scandinavian, Danish and German working groups as well as the written deliberations of the North Atlantic Arctic Gateway Detailed Planning Group (Ruddiman *et al.* 1991) and of the Joides advisory structure. Their contributions are gratefully acknowledged.

References

Barry, R. G. 1989 The present climate of the Arctic Ocean and possible past and future states. In *The Arctic seas: climatology, oceanography, geology and biology* (ed. Y. Herman), pp. 1–46. New York: van Nostrand Reinhold.

Broecker, W. S. 1991 The great ocean conveyor. *Oceanography* **4**, 79–89.

Carter, L. D., Brigham-Grette, J., Marinkovich, L., Pease, V. L. & Hillhouse, J. W. 1986 Late Cenozoic Arctic Ocean sea ice and terrestrial palaeoclimate. *Geology* **14**, 675–678.

CIA 1978 Polar regions – atlas. Central Intelligence Agency. (Reprinted 1981.)

Clark, D. L. 1982 Origin, nature, and world climate effect of Arctic Ocean ice-cover. *Nature, Lond.* **300**, 321–325.

Creager, J. S. *et al.* 1973 *Init. Reports* DSDP **19**. Washington, DC: US Government Printing Office.

Einarsson, T. & Albertsson, K. J. 1988 The glacial history of Iceland during the past three million years. In *The past three million years: evolution of climatic variability in the North Atlantic region* (ed. N. J. Shackleton, R. G. West & D. Q. Bowern), pp. 227–234. Cambridge University Press.

Eldholm, O. *et al.* 1987 *Proc.* ODP *Init. Rep.* **104**. College Station, TX: Ocean Drilling Program.

Eldholm, O. & Thomas, E. 1993 Environmental impact of volcanic margin formation. *Earth planet. Sci. Lett.* **117**, 319–329.

Funder, S., Abrahamsen, N., Bennike, O. & Feyling-Hanssen, R. W. 1985 Forested Arctic: evidence from North Greenland. *Geology* **13**, 542–546.

Henrich, R. 1992 Beckenanalyse des Europäischen Nordmeeres: Pelagische und glaziomarine Sedimentflüsse im Zeitraum 2.6 Ma bis rezent. *Habil.-Thesis, Fac. Nat. Sci.* Kiel University.

Herman, Y. & Hopkins, D. M. 1980 Arctic Ocean climate in late Cenozoic time. *Science, Wash.* **209**, 557–562.

Hinz, K., Eldholm, O., Block, M. & Skogseid, J. 1993 Evolution of North Atlantic volcanic continental margins. In *Petroleum Geology of Northwest Europe: Proc. 4th Conf.* (ed. J. R. Parker), pp. 901–913. London: Geological Society.

Jansen, E. & Sjøholm, J. 1991 Reconstruction of glaciation over the past 6 million years from ice-borne deposits in the Norwegian Sea. *Nature, Lond.* **349**, 600–604.

Jansen, E., Bleil, U., Henrich, R., Kringstad, L. & Slettemark, B. 1988 Palaeoenvironmental changes in the Norwegian Sea and Northeast Atlantic during the last 2.8 m.y.: Deep-Sea Drilling Project/Ocean Drilling Program Sites 610, 642, 643 and 644. *Palaeoceanography* **3**, 563–581.

Koltermann, K. P. 1987 Tiefenzirkulation der Grönland-See als Folge des thermohalinen Systems des Europäischen Nordmeeres. Ph.D. thesis. Faculty of Natural Sciences, Hamburg University.

Kulm, L. D., *et al.* 1973 *Init. Reports* DSDP **18**. Washington, DC: US Government Printing Office.

Meincke, J. 1983 The modern current regime across the Greenland–Scotland Ridge. In *Structure and development of the Greenland–Scotland Ridge: new methods and concepts* (ed. M. H. P. Bott, S. Saxov, M. Talwani & J. Thiede), pp. 673–650. New York: Plenum.

Mudie, P. J. & Helgason, J. 1983 Palynological evidence for Miocene climatic cooling in eastern Iceland about 9.8 Myr ago. *Nature, Lond.* **303**, 689–692.

Myhre, A. M. *et al.* 1995 *Proc.* ODP *Init. Rep.* **151**. College Station, TX: Ocean Drilling Program.

Nelson, R. E. & Carter, L. D. 1985 Pollen analysis of a late Pliocene and early Pleistocene section from the Gubik Formation of Arctic Alaska. *Quat. Res.* **24**, 295–306.

Quadfasel, D., Gascard, J.-C. & Koltermann, K.-P. 1987 Large-scale oceanography in Fram Strait during the 1984 Marginal Ice Zone Experiment. *J. Geophys. Res.* **92** (C 7), 6719–6728.

Raymo, M. E., Rind, D. E. & Ruddiman, W. F. 1990 Climatic effects of reduced Arctic sea ice limits in the GISS II General Circulation Model. *Palaeoceanography* **5**, 367–382.

Rea, D. K., Basov, L. A. & Janecek, T. R. 1993a North Pacific Transect. *JOIDES J.* **19**. 21–28.

Rea, D. K., *et al.* 1993b *Proc.* ODP *Init. Rep.* **145**. College Station, TX: Ocean Drilling Program.

Repenning, C. A., Brouwers, E. M., Carter, L. D., Marincovich, L. Jr & Ager, T. A. 1987 The Beringian anchestry of Phenacomys Criceridae) and the beginning of the modern Arctic Ocean borderland biota. *US Geol. Surv. Bull.* **1687**, 1–28.

Ruddiman, W. F. & JOIDES NAAG-DPG 1991 North Altantic–Arctic Gateways. *Joides J.* **17**, 38–50.

Ruddiman, W. F. & Raymo, M. E. 1988 Northern Hemisphere climatic régimes during the past 3 Ma: possible tectonic connections. *Phil. Trans. R. Soc. Lond.* B **318**, 411–430.

Ruddiman, W. F., Raymo, M. & McIntyre, A. 1986 Matuyama 41 000-year cycles: North Atlantic Ocean and Northern Hemisphere ice sheets. *Earth planet. Sci. Lett.* **80**, 117–129.

Talwani, M., *et al.* 1976 DSDP *Leg 38 Init. Rep.* **38**. Washington, DC: US Government Printing Office.

Thiede, J., Eldholm, O. & Taylor, E. 1989 Variability of Cenozoic Norwegian Greenland Sea palaeoceanography and Northern Hemisphere palaeoclimate. In *Proc.* ODP, *Sci. Res.* (ed. O. Eldholm *et al.*), **104**. College Station, TX: Ocean Drilling Program, pp. 1067–1118.

Thiede, J. & NAD Science Committee 1992 The Arctic Ocean record: key to global change (initial science plan). *Polarforschung* **61**, 1–102.

Warnke, D. A. & Hansen, M. E. 1977 Sediments of glacial origins in the area of DSDP leg 38 (Norwegian Greenland seas): Preliminary results from Sites 336 and 344. *Naturforsch. Ges. Freib. Breisgau Ber.* **67**, 371–392.

Wolf, T. C. W. & Thiede, J. 1991 History of terrigenous sedimentation during the past 10 m.y. in the North Atlantic (ODP Legs 104, 105, and DSDP 81). *Mar. Geol.* **101**, 83–102.

Index

Printed and bound by CPI Group (UK) Ltd, Croydon, CR0 4YY

24/10/2024

01778287-0002